COMPREHENSIVE BIOCHEMISTRY

ELSEVIER SCIENTIFIC PUBLISHING COMPANY

335 Jan van Galenstraat, P.O. Box 211, Amsterdam, The Netherlands

AMERICAN ELSEVIER PUBLISHING COMPANY, INC.

52 Vanderbilt Avenue, New York, N.Y. 10017

Library of Congress Card Number 62–10359
ISBN 0-444-41281-6

With 22 illustrations and 12 tables

COMPREHENSIVE BIOCHEMISTRY

COMPREHENSIVE BIOCHEMISTRY

SECTION I (VOLUMES 1–4)
PHYSICO-CHEMICAL AND ORGANIC ASPECTS OF BIOCHEMISTRY

SECTION II (VOLUMES 5–11)
CHEMISTRY OF BIOLOGICAL COMPOUNDS

SECTION III (VOLUMES 12–16)
BIOCHEMICAL REACTION MECHANISMS

SECTION IV (VOLUMES 17–21)
METABOLISM

SECTION V (VOLUMES 22–29)
CHEMICAL BIOLOGY

SECTION VI (VOLUMES 30–33)
A HISTORY OF BIOCHEMISTRY

COMPREHENSIVE BIOCHEMISTRY

EDITED BY

MARCEL FLORKIN

Professor of Biochemistry, University of Liège (Belgium)

AND

ELMER H. STOTZ

Professor of Biochemistry, University of Rochester, School of Medicine and Dentistry, Rochester, N.Y. (U.S.A.)

VOLUME 25

REGULATORY FUNCTIONS – MECHANISMS OF HORMONE ACTION

ELSEVIER SCIENTIFIC PUBLISHING COMPANY

AMSTERDAM · LONDON · NEW YORK

1975

CONTRIBUTORS TO THIS VOLUME

L. BIRNBAUMER, Ph.D.
Department of Physiology, Northwestern University Medical Center, Chicago, Ill. 60611 (U.S.A.)

T. BRAUN, M.D., Ph.D.
Department of Physiology, Northwestern University Medical Center, Chicago, Ill. 60611 (U.S.A.)

D. DOENECKE, Dr.
Institut für Physiologische Chemie I, Institutsgruppe Lahnberge der Medizinischen Fakultät der Philipps-Universität, D-355 Marburg (Lahn) (Bundesrepublik Deutschland)

P. KARLSON, Prof.Dr.Dr.h.c.
Institut für Physiologische Chemie, Institutsgruppe Lahnberge der Medizinischen Fakultät der Philipps-Universität, D-355 Marburg (Lahn) (Bundesrepublik Deutschland)

C. E. SEKERIS, Dr.
Institut für Physiologische Chemie, Institutsgruppe Lahnberge der Medizinischen Fakultät der Philipps-Universität, D-355 Marburg (Lahn) (Bundesrepublik Deutschland)

GENERAL PREFACE

The Editors are keenly aware that the literature of Biochemistry is already very large, in fact so widespread that it is increasingly difficult to assemble the most pertinent material in a given area. Beyond the ordinary textbook the subject matter of the rapidly expanding knowledge of biochemistry is spread among innumerable journals, monographs, and series of reviews. The Editors believe that there is a real place for an advanced treatise in biochemistry which assembles the principal areas of the subject in a single set of books.

It would be ideal if an individual or small group of biochemists could produce such an advanced treatise, and within the time to keep reasonably abreast of rapid advances, but this is at least difficult if not impossible. Instead, the Editors with the advice of the Advisory Board, have assembled what they consider the best possible sequence of chapters written by competent authors; they must take the responsibility for inevitable gaps of subject matter and duplication which may result from this procedure.

Most evident to the modern biochemist, apart from the body of knowledge of the chemistry and metabolism of biological substances, is the extent to which he must draw from recent concepts of physical and organic chemistry, and in turn project into the vast field of biology. Thus in the organization of Comprehensive Biochemistry, the middle three sections, Chemistry of Biological Compounds, Biochemical Reaction Mechanisms, and Metabolism may be considered classical biochemistry, while the first and last sections provide selected material on the origins and projections of the subject.

It is hoped that sub-division of the sections into bound volumes will not only be convenient, but will find favour among students concerned with specialized areas, and will permit easier future revisions of the individual volumes. Toward the latter end particularly, the Editors will welcome all comments in their effort to produce a useful and efficient source of biochemical knowledge.

M. FLORKIN

E. H. STOTZ

Liège/Rochester

PREFACE TO SECTION V

(VOLUMES 22–29)

After Section IV (*Metabolism*), Section V is devoted to a number of topics which, in an earlier stage of development, were primarily descriptive and included in the field of Biology, but which have been rapidly brought to study at the molecular level. "*Comprehensive Biochemistry*", with its chemical approach to the understanding of the phenomena of life, started with a first section devoted to certain aspects of organic and physical chemistry, aspects considered pertinent to the interpretation of biochemical techniques and to the chemistry of biological compounds and mechanisms. Section II has dealt with the organic and physical chemistry of the major organic constituents of living material, including a treatment of the important biological high polymers, and including sections on their shape and physical properties. Section III is devoted primarily to selected examples from modern enzymology in which advances in reaction mechanisms have been accomplished. After the treatment of Metabolism in the volumes of Section IV, "*Comprehensive Biochemistry*", in Section V, projects into the vast fields of Biology and deals with a number of aspects which have been attacked by biochemists and biophysicists in their endeavour to bring the whole field of life to a molecular level. Besides the chapters often grouped under the heading of molecular biology, Section V also deals with modern aspects of bioenergetics, immunochemistry, photobiology and finally reaches a consideration of the molecular phenomena that underlie the evolution of organisms.

M. FLORKIN

E. H. STOTZ

Liège/Rochester

CONTENTS

REGULATORY FUNCTIONS—MECHANISMS OF HORMONE ACTION

Chapter I. Intracellular Mechanisms of Hormone Action
by P. KARLSON, D. DOENECKE AND E. SEKERIS

Chapter II. Hormone-Sensitive Adenyl Cyclase Systems: Properties and Function

by THEODOR BRAUN AND LUTZ BIRNBAUMER

COMPREHENSIVE BIOCHEMISTRY

Section I — Physico-Chemical and Organic Aspects of Biochemistry
Volume 1. Atomic and molecular structure
Volume 2. Organic and physical chemistry
Volume 3. Methods for the study of molecules
Volume 4. Separation methods

Section II — Chemistry of Biological Compounds
Volume 5. Carbohydrates
Volume 6. Lipids — Amino acids and related compounds
Volume 7. Proteins (Part 1)
Volume 8. Proteins (Part 2) and nucleic acids
Volume 9. Pyrrole pigments, isoprenoid compounds, phenolic plant constituents
Volume 10. Sterols, bile acids and steroids
Volume 11. Water-soluble vitamins, hormones, antibiotics

Section III — Biochemical Reaction Mechanisms
Volume 12. Enzymes — general considerations
Volume 13. (third edition). Enzyme nomenclature (1972)
Volume 14. Biological oxidations
Volume 15. Group-transfer reactions
Volume 16. Hydrolytic reactions; cobamide and biotin coenzymes

Section IV — Metabolism
Volume 17. Carbohydrate metabolism
Volume 18. Lipid metabolism
Volume 19. Metabolism of amino acids, proteins, purines, and pyrimidines
Volume 20. Metabolism of cyclic compounds
Volume 21. Metabolism of vitamins and trace elements

Chapter I

Intracellular Mechanisms of Hormone Action

P. KARLSON, D. DOENECKE and C. E. SEKERIS

Institute of Physiological Chemistry, Philipps University Marburg (Germany)

The present review deals only with the intracellular activity of hormones. It does not consider mechanisms such as the control of cell permeability at the level of the cell membrane, presumably exerted by some hormones; nor is it concerned with the contribution made by hormones *via* membrane-bound adenylate cyclase to the production of cyclic AMP. The latter subject is covered in Chapter II of this Volume.

Though the interaction of hormones with both cytoplasmic and nuclear components will be discussed, emphasis will be placed on the effect of hormones on nuclear material, especially chromatin. The reason for this will become clear during our discussion.

1. Historical introduction

The gross physiological effect of a hormone is usually detected first; indeed such detection is the prerequisite for the description·of a new hormone. Understanding of the molecular action of hormones has come about very slowly. As a chemical compound, a hormone must interact at the proper site of a target cell with some other chemical compound; it is only as a result of this interaction, possibly mediated by many other biochemical or physiological events, that the final "physiological effect" is brought about.

In the past, several types of primary interaction on the part of a hormone with a cell component have been put forward by way of explanation of hormone action (Review: Karlson[1,2,2a]). In the early 1940's the idea that hormones could interact with enzymes in a manner similar to vitamins (*i.e.* as "cofactors") was quite popular (Green[3]). Until then, vitamins had been identified as parts of coenzymes; that hormones might work in a similar fashion was a novel and fascinating idea. However, in spite of many efforts to demonstrate the interaction of hormones and enzymes along these lines, no clear-cut example was found (Reviews: Hechter[4], Dirscherl[5]) and the idea was gradually abandoned.

For a brief while the "coenzyme concept" was revived as a result of the observation that steroids such as oestradiol stimulate transhydrogenation between NADPH and NAD$^+$ or $vice versa$, $i.e.$ between the NADP$^+$/NADPH and the NAD$^+$/NADH-system (Villee et $al.$[6]; Talalay and Williams-Ashman[7]), but this likewise failed to explain the physiological action of oestrogens in biochemical terms.

It should be kept in mind, however, that at the time the concept of allosteric modification of enzyme activity did not yet exist. This concept, of course, is quite different from that ascribing to the hormone the role of a coenzyme. On the other hand, if there were remarkable changes in enzyme activity through allosteric effects, they should have been detected in the course of the investigations mentioned above. But only one case—the steroid effects on glutamate dehydrogenase studied by Tomkins and co-workers[8]—attracted attention. At first, these authors tried to explain the steroid hormone action on this basis. The effect lacks specificity, however, and for this and other reasons this explanation had to be abandoned. In light of the numerous studies on interaction between steroid hormones and enzymes, it seems unlikely that these hormones act as allosteric effectors of enzymes.

The notion that hormones act as regulators of gene activity (Karlson[2]) arose from the observation that the insect hormone ecdysone could induce, in larvae of $Chironomus$, the puffing of two specific bands of salivary-gland giant chromosomes[9]. The following scheme adopted from Karlson[2] (Fig. 1) outlines what is meant.

The puff is made visible by staining technique and can be observed directly in the light microscope. It shows a looser DNA structure. The staining properties are due to proteins present in the puff. Pelling[10] has shown, and this has later been confirmed by other authors, that puffs are the site of active RNA synthesis. On the basis of these observations, the following chain of events can be foreseen.

As a result of the primary action of the hormone, the puff is formed and RNA is actively synthesized in the puff. The RNA is regarded as messenger

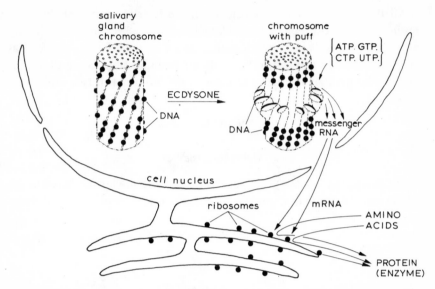

Fig. 1. Mechanism of action of ecdysone. The hormone acts first on the DNA producing a puff, which is shown on the right as an unwound region. In the puff, RNA is synthesized from precursors. This RNA is believed to be transferred to the cytoplasm and attached to the ribosomes. As "messenger-RNA", it carries the information about the amino acid sequence. According to this information, the specific protein is synthesized on the ribosome from activated amino acids. The whole chain of events explains the formation of specific proteins (*e.g.*, certain enzymes) as the response of the target cell to a hormone. (From Karlson[2])

RNA. (Ribosomal RNA is made in the nucleolus, to which the site of the puff is not related.) This messenger RNA is then transferred to the cytoplasm where it is translated into the polypeptide chain of a protein, possibly an enzyme or a protein of other physiological significance.

A number of biochemical conclusions can be drawn from this hypothesis. (*1*) The hormone must be present in the cell nucleus since that is its site of action. (*2*) RNA synthesis must be stimulated by the hormone. It should be detectable by biochemical methods, *e.g.* by the incorporation of isotopes into RNA. (*3*) Inhibition of RNA synthesis by metabolic inhibitors such as actinomycin must abolish the effect of the hormone, since RNA synthesis (transcription) would no longer occur. (*4*) The ultimate result of the events in the nucleus would be the production of messenger RNA. It would be transferred to the cytoplasm and translated into a protein. This might be an enzyme that can be detected by enzyme assay. This last step could also be inhibited by a proper metabolic inhibitor, *e.g.* cycloheximide.

References p. 54

All these biochemical conclusions can be tested experimentally. As a result, this new concept has considerably influenced research on primary molecular processes in hormone action, and a large number of studies has been published on this subject[11-16]. It will be impossible to list all papers, but it is hoped that the views of most of the authors will be properly represented.

2. Steroid hormones

(a) General considerations

Work during the last ten years has shown striking similarities between the mechanisms of action of steroid hormones. The first intracellular event seems

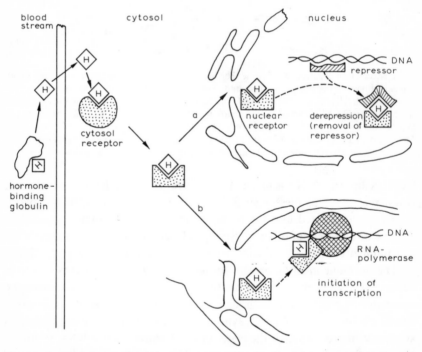

Fig. 2. Role of receptor proteins in hormone action. The hormone (H) is carried in the blood by a globulin. When it enters the cell, it is bound to the cytosol receptor which in turn is converted into the form capable of entering the nucleus. The control of transcription can either be negative (arrow 'a'), *i.e.* by a mechanism of derepression, visualized as removal of a repressor protein, or positive (arrow 'b'), visualized as positive cooperation of hormone-receptor with RNA polymerase to start transcription at this specific operon.

to be a binding to a cytoplasmic protein generally referred to as "steroid receptor". This hormone–receptor complex is then transferred to the cell nucleus; in this process, the hormone receptor is altered. Distinct nuclear receptor proteins may also exist, contributing to the formation of the intranuclear hormone–receptor complexes.

The next well-documented step is stimulation of nuclear RNA synthesis at the chromatin level. The RNA produced is in part the heterogeneous nuclear RNA which very probably contains the specific messenger RNA sequences. Indeed, in some cases, the formation of mRNA for distinct proteins has been well documented (see Section 2c, p. 17). Later, the formation of nucleolar RNA is also enhanced and new ribosomes are formed in the cytoplasm.

Several hours after hormone application, an increase in protein synthesis is observed. This effect can be studied either in terms of the incorporation of labelled precursors into cellular protein or in terms of the induction of specific enzymes or hormone-dependent proteins. Protein induction has often been studied independently of RNA synthesis, and it has been disputed whether this is the result of transcriptional or translational control. This question is discussed in detail on p. 42.

On the basis of these events, a diagram showing the action of steroid hormones can be drawn (Fig. 2). In it, the control of protein induction is shown as transcriptional control, occurring at the level of mRNA synthesis.

(b) Receptor proteins for steroid hormones

The existence in target tissues of intracellular protein molecules that bind steroid hormones has been demonstrated in recent years[15]. "Receptors" are virtually absent from non-target tissues. Receptor molecules are found in the cytosol and in the nucleus. In the cytosol, receptors usually exist in the form of aggregates sedimenting with S values of over 6 which can be dissociated in the presence of high salt concentrations to molecules sedimenting with S values of 3–5. 4–5 S receptors can be extracted from nuclei with high salt concentrations. Currently they are believed to be modified cytosol receptors, although the existence of specific nuclear receptors cannot be ruled out. A survey of the properties of "receptor proteins" is given in Table I. It may be assumed that receptor proteins play an important role in the action mechanism of hormones.

The role of receptor proteins in hormone action is shown in Fig. 2. In the

TABLE I

RECEPTOR PROTEINS FOR STEROID HORMONES

Hormone	Tissue	Nucleus	Cytosol	Designation mol. wt. or (S)	Separation	Characteristics, Specificity	Ref.
Oestrogens	uterus	+		(5 S)	Density gradient centrifugation	Is formed after incubation of nuclei with the cytosol–receptor–hormone complex at 37° or by heat transformation of the 4 S cytosol receptor.	40 41 58
			+	(8 S)	Density gradient (in low salt)	4 S and 8 S are reversibly interchanged depending on ionic strength of medium. By heating or after ammonium sulphate precipitation they are transformed to a 5 S receptor similar or identical to the nuclear 5 S protein.	46 49
			+	(4 S)	(in high salt)		
Progesterone	uterus		+	(4 S)	Sucrose gradient centrifugation	Similar to transcortin.	68
			+	(6–7 S)	Sucrose gradient centrifugation (in low salt)		
			+	(4 S)	(in high salt)	Specific for progesterone. Induced by oestrogens (according to Ref. 70, 7.6 S in low salt and 5.5–6.0 S in high salt).	69, 70 69, 70
	chick oviduct	+	+ A + B	(4 S) (8 S)	Agarose A-0.5 filtration (in low salt)	A is specific for progesterone.	72
				(3.7 S)	(in high salt)	B is similar to transcortin.	72

TABLE I (continued)

Hormone	Tissue				S value / MW	Method	Description	Ref.
Androgens	prostate	+			(8 S)	Sucrose gradient centrifugation	Is formed after incubation of nuclei with cytosol receptor–androgen complex II.	77 78
						DNA-cellulose chromatography		80
			+	complex I (3.5) complex II (3.5)		Sephadex G-200 filtration	Receptors preferentially bind dihydrotestosterone; to a much lesser extent, testosterone.	77 78
	rat testis		+		(8 S)	Sucrose gradient	Same as I + II, S value depends on ionic strength of medium.	79
			+		(4 S)	Sucrose gradient	Carrier of androgen from testis to epididymis.	84
	ductus deferens tumor		+		(5.7 S)	Sucrose gradient	Involved in nuclear translocation of testosterone to a nuclear acceptor.	83
			+		(5.1 S)	Sucrose gradient		
Aldosterone	toad bladder		+		(3 S)		Soluble chromatin-bound.	87– 91
			+		(4 S)			
			+		(5 S)			
	kidney		+		(8.5 S/4 S) (low salt) (4.5 S) (high salt)	Glycerol gradient	Temperature-dependent transfer to nuclear acceptor sites.	22
Glucocorticoids	liver	+			(7 S)	Sucrose gradient centrifugation		35
			+	A	51 000	DEAE-cellulose	Binds natural glucocorticoids.	92
			+	B	64 000	DEAE-cellulose	Similar to transcortin.	92
			+	G	61 000 (in high salt)	Sephadex G-150	Binds natural and synthetic glucocorticoids.	93 95
			+	I	31 000	DEAE–Sephadex	Binds anionic metabolites of cortisol.	97 98

References p. 54

TABLE I (continued)

Hormone	Tissue	Nucleus	Cytosol	Desig-nation	mol. wt. or (S)	Separation	Characteristics, specificity	Ref.
			+	II		DEAE–Sephadex	Binds cortisol (similar to G).	98
			+	III		DEAE–Sephadex	Binds anionic metabolites.	98
			+	IV		DEAE–Sephadex	Binds cortisol with low, progesterone and testosterone with higher affinity (similar to B).	98
	thymus	+			(7 S)	DEAE–cellulose, Sephadex G-200	Binds natural and synthetic steroids; in presence of high salts dissociates partially to 4 S molecules.	101 101a
			+		(4 S)	DEAE–cellulose	Mixture of two receptors, one similar to transcortin and one similar to the nuclear receptor.	101 102
								103–106
	lymphoma hepatoma cells		+		(4 S)	Sucrose gradient centrifugation (high ionic strength)	Activation of steroid–receptor complex by dilution, ionic strength or temperature allows nuclear binding.	111, 112
			+		(7 S)	Sucrose gradient centrifugation (low ionic strength)		
					(4 S)			117
Ecdysone	larvae lymph					Electrophoresis	Proteins similar to transport proteins of serum.	118
	salivary glands	+				Sucrose gradient centrifugation (high salt)	(8 S) at low salt concentration.	120
	hepatopancreas		+ +		(4 S)	Sucrose gradient centrifugation	Binds an ecdysone metabolite.	123

blood stream, the hormone is bound either specifically to a plasma protein (*e.g.* transcortin for cortisol) or non-specifically to albumin. In the target tissue, the hormone is taken up by the target cell either by diffusion or by some other, as yet unknown transport mechanism. Within the cell, the steroid is bound to the cytoplasmic receptor. The dissociation constant of the steroid–receptor complex is lower by one to two orders of magnitude as compared to the transport complex in blood plasma. Thus, even in the absence of an active transport system, the hormone will accumulate within the cell.

The cytoplasmic receptor–hormone complex is believed to undergo a transition to become the "nuclear receptor" which in the nucleus interacts with the chromatin to induce RNA synthesis. The mechanism of this inter-action and the possible role of nuclear-receptor molecules in this process are still unknown. In Fig. 2, the possibility of positive as well as negative control is indicated.

Since the dissociation constants decrease considerably from step to step, which is one way of saying that the binding becomes stronger, the hormone will eventually accumulate in the nucleus.

This has indeed been found in the case of oestradiol in autoradiographic studies. They show that oestradiol in target tissues is mainly concentrated in the cell nucleus (Stumpf[17]).

It should be pointed out that this is a dead-end road. There is no way to allow the diffusion of the hormone back into the circulation. However, it is known that the hormone does not persist forever in its target tissue. There must be a mechanism for removal of the hormone. One possibility would be the hormone's metabolism; however, it has been shown that oestradiol is not transformed or metabolized in target tissue. How the hormone is eventually removed is still unknown; this would be an interesting subject for further studies.

Receptor proteins are present in appreciable amounts only in cells of target tissues. This is believed to be the molecular basis for a "target tissue". If the binding of the steroid to a special receptor molecule plays an important part in its mechanism of action, then obviously only those tissues containing the receptor will react to the hormone and behave as target tissue.

The following findings point to the key role of receptor proteins in hormone action: (*1*) Hormone receptors are present in appreciable amounts in target tissues only. (*2*) Administration of hormone analogues which bind to receptor proteins decreases the uptake of hormone by the tissue and abolishes tissue response to the hormone[18–22]. (*3*) Certain tumours which are sensitive to hormones have high titres of receptors to these hormones, whereas resistant tumours have a markedly reduced receptor titre[23-28]. (4) The saturation of the receptors after hormone administration is markedly parallel to the degree of the hormonal effect, e.g. enzyme induction[29,30]. (5) Receptor proteins are present in low concentrations during periods when the hormone elicits no response[31,32]. (6) Some direct effects of hormone receptors on in vitro RNA synthesis have been reported[33-38].

(i) Oestradiol receptors

Receptors for oestradiol have first been observed by Jensen and co-workers[39-43].

The most important tool in these studies was highly labelled oestradiol that could be traced in the tissues even when administered in physiological doses. It was shown that oestradiol becomes bound to protein in target tissues, and that the hormone is concentrated in target cells. The concentration of the hormone in non-target tissues is the same as in the blood, *i.e.* there is a free equilibrium between tissue and blood. Peck and co-workers[44] measured the rate of uptake of oestradiol by a non-target tissue, the diaphragm, which possesses no oestrogen receptor. They were able to show that the rate was the same as for uterine tissue and concluded that the receptor molecule is not involved in the transport or uptake process but in the *retention* of oestrogen in the uterus.

Two sites of oestrogen-binding have been detected in target cells, one in the cytoplasm and one in the nucleus (Maurer and Chalkley[45]). In the cytosol a protein sedimenting with an S value of 8 in sucrose gradients has been detected[46,47]. In a medium of high ionic strength, the protein reversibly dissociates to a 4 S protein, presumably the subunit form of the 8 S aggregate[48-50]. This dependence of receptor aggregation on salt concentration was clearly demonstrated by Stancel *et al.*[51] by incubating the receptor at different salt concentrations. They concluded from their experimental results that *in vivo* the oestrogen–receptor complex exists as a 3.8–4.8 S form.

Besides this influence of the ionic environment, the probable role of proteases has been demonstrated by Notides *et al.*[52]. These authors reproducibly converted 8 S receptors from rat uterus into a 4.5 S oestrogen–receptor complex using a human uterine protease.

From endometrium nuclei a protein–oestradiol complex can be isolated by high salt extraction sedimenting at approximately 5 S. *In vivo*, and in the absence of oestradiol there is no evidence for the presence of the 5 S protein in the nucleus. Incubation of the nuclei with oestradiol alone does not lead to formation of the 5-S complex whereas addition of the cytosol receptor to the nuclear suspension leads immediately to a temperature-dependent translocation of the receptor into the nucleus and to formation of the 5-S complex (Jensen et al.[40]). The 5-S complex can also be formed by heating the cytosol 4-S oestradiol complex in the absence of the nuclei, showing an optimum of formation at pH 6.5–8.5, or after precipitation with ammonium sulphate. In the presence of EDTA the transformation does not take place.

Several laboratories have isolated and purified the oestradiol–receptor proteins[53–56]. The cytoplasmic receptor has been obtained in nearly pure form (Jensen et al.[53]). The nuclear receptor has also been purified to a certain extent[57,58]. According to Mohla et al.[59] and DeSombre et al.[60], the role of the 5-S nuclear receptor (transformed 4-S cytoplasmic receptor) is to stimulate transcription. The hormone is needed only for the translocation of the receptor and does not take part in the molecular events on the chromatin. At least this appears to be the case on the basis of stimulation of the transcription of isolated uterine nuclei using a heat-transformed or ammonium sulphate-precipitated receptor in the absence of oestradiol (Mohla et al.[59]). Further, Jensen and co-workers have shown that the transition of the 4-S receptor to the 5-S form can be prevented by "stabilizing" the receptor in the presence of Ca^{2+} and salt. This stabilized receptor, although it does bind oestradiol, cannot affect the RNA synthetic capacity of isolated endometrium nuclei.

Treatment of cytosol with trypsin or calcium ions results in the inability of the 4-S receptor complex to bind to DNA (Andre and Rochefort[61]). On the other hand, Bresciani et al.[28] postulate a Ca^{2+}-dependent receptor-transforming factor (RTF), which produces a 4.5-S receptor fragment that enters the nucleus and there stimulates transcription after binding to chromosomal proteins. According to Puca et al.[58], these nuclear binding sites for oestrogen–receptor complexes are exclusively in the basic nuclear-protein fraction; receptors do not bind to DNA itself.

The mechanism of nuclear binding of receptor complexes has been studied with isolated chromatin and purified DNA. The initial discovery by Shyamala-Harris[62] that DNA is involved in the nuclear binding of the hormone–receptor complex was corroborated by affinity chromatography studies with column-bound DNA[63–65]. In these studies, hormone–receptor complexes could be eluted from DNA by buffers of the ionic strength used to elute in vivo bound nuclear receptors. Even a certain specificity of the

4-S receptor for host DNA was described by Clemens and Kleinsmith[63].

The role of non-histone proteins, which has been shown by O'Malley and co-workers[66] to be a major one in nuclear-receptor binding, has mainly been studied in the progesterone-receptor system (see below).

Oestradiol receptors have also been detected in other target tissues such as the hypophysis, oestrogen-dependent tumours and chick oviduct. The characteristics of these molecules are very similar to those described for rat uterus.

Recently an oestrogen receptor confined to the cell nucleus and having a very high affinity $(K_d = 10^{-14}\ M)$ has been described by Baulieu and co-workers[67] in calf endometrium nuclei.

(ii) Progesterone receptors

A progesterone-binding protein has been isolated from the cytosol of rat uterus; it has the characteristics of the corticoid-binding globulin of serum (Milgrom and Baulieu[68]).

A specific cytosol receptor for progesterone has been found in the guinea pig showing an S value of 6.7–7.6 S in low salt and of 3–3.5 S in the presence of 0.5 M KCl (Milgrom et al.[69]; Faber et al.[70]). The K_a is $5 \cdot 10^8\ M^{-1}$. By extracting nuclei with 0.3 M KCl a receptor with an S value of approximately 4 could be detected on sucrose gradients. The amount of the cytosol receptor increases after oestradiol administration and is diminished by progesterone itself[71]. Thus a double control of the progesterone receptor concentration exists, a positive one through oestrogen and a negative one by progesterone.

From the cytosol of chick oviduct, two fractions binding progesterone have been separated by DEAE-cellulose chromatography[72]. Fraction A has been purified approximately 800-, fraction B 3000-fold. Both sediment at 4 S. They are rapidly taken up by isolated oviduct nuclei[73]. Fraction A binds preferentially to DNA whereas fraction B has a high affinity to oviduct chromatin but not to chromatin from non-target tissues. The two fractions seem to be subunits of one protein. According to Steggles et al.[74], chromatin from target tissues contains "acceptor sites" specific for the cytosol receptors. The experimental data on which this concept is based have recently been reviewed by O'Malley and Means[66]. They conclude that a high affinity reaction may occur between subunit A and chromatin DNA, whereas subunit B of the intact receptor binds to a nonhistone acceptor protein. The role of these proteins in the specificity of nuclear receptor binding was demonstrated in chromatin-reconstitution experiments, in which "hybrid"-

chromatins, reconstituted out of components from target and non-target tissues, were assayed (Spelsberg et al.[75]). The important role of nonhistone proteins in nuclear receptor binding was also demonstrated by Chatkoff and Julian[76] in studies with rabbit uterine chromatin.

(iii) Androgens

Androgen receptors have been detected both in the cytosol and the nuclei of target organs. Most of the work done deals with receptors present in the prostate gland. The cytosol receptor shows an S value of 3.5 S (ref. 77) and has been separated by DEAE-cellulose chromatography into fractions I and II (ref. 78). Mainwaring[79] has also described a cytosol receptor in prostate sedimenting at 8 S. It appears that, as in the case of the other receptors, the S value reflects the ionic strength of the medium used in the experiments and not the true molecular weight of the protein. Recently, Mainwaring and Irving[80] isolated a receptor protein from the same tissue by DNA cellulose chromatography and subsequent isoelectric focussing. This protein was able to transfer testosterone to the nucleus.

From the nucleus, a 3-S receptor can be extracted. The formation of the 3-S receptor after incubation of prostate nuclei in the presence of the 3.5-S cytosol receptor is reminiscent of conditions governing the case of oestrogens and endometrium nuclei.

A 3.5-S nuclear receptor has been described by Blaquier and Calandra[81] in rat epididymal nuclei.

These androgen-receptor proteins preferentially bind dihydrotestosterone and, to a much smaller extent, testosterone. Dihydrotestosterone appears to be the active form of the androgens. Competing agents in the binding are oestradiol, progesterone and cyproterone.

Binding studies in androgen-insensitive mutants have been performed with kidney cytoplasmic receptor preparations from mice carrying the testicular feminization (Tfm) defect. Gehring et al.[82] demonstrated very low level in cytoplasmic hormone uptake as well as in nuclear binding.

Further descriptions of cytoplasmic androgen receptors include ductus deferens tumours[83], rat testis[84], rat levator ani muscle[85] and uterine tissues[86], each reflecting the metabolic requirements or hormone sensitivities of the respective tissues.

(iv) Aldosterone-binding proteins

Being a convenient target tissue for the action of aldosterone, the toad bladder has been studied in many laboratories. It has been shown that aldosterone is concentrated. Two proteins are apparently involved, a cytoplasmic binding protein and one confined to the nucleus[87-91]. The same or at least a similar protein is also present in the mucosa of the gut. Its function in this case is not clear since the sodium transport in the gut is not under the control of aldosterone. Similar receptor proteins have been detected in the kidney (Marver et al.[22]). They can be occupied by the anti-mineralocorticoid Spirolactone, but this complex is unable to bind to chromatin acceptor sites. The authors conclude from their results that the receptor exists in both an active and an inactive configuration; the hormone binds to the active, the antagonist to the inactive form, thus preventing the receptor from being transferred to nuclear sites.

(v) Glucocorticosteroid receptors

Glucocorticosteroid binding proteins have been detected in many of the target tissues examined, including the liver, lung thymus, lymph cells, fibroblasts and hepatoma cells in culture.

In rat liver three binding proteins have been found, A, B, and G. They can be separated by DEAE-cellulose chromatography and Sephadex G-150 gel filtration (Beato et al.[92,93]; Koblinsky et al.[94]; Sekeris and Schmid[95]). All three proteins bind cortisol and corticosterone with high affinity and specificity. Protein B has similar binding characteristics to serum trans-cortin. Neither A nor B is able to bind synthetic glucocorticosteroids such as dexamethasone whereas binder G does do so. The molecular weight of binder A is approximately 51 000, that of binder B 64 000. In the presence of high salt concentrations binder G shows a molecular weight of 66 000 but tends to form aggregates sedimenting at 7–9 S in low ionic strength media. The following evidence favours the assumption that binder G is the principal hepatic glucocorticoid receptor protein: (a) it binds the synthetic glucocorticosteroid, dexamethasone, while the other two binders do not (Koblinsky et al.[94]); (b) the saturation of binder G with cortisol after *in vivo* application correlates well with the degree of enzyme induction (Beato et al.[30]); (c) receptor G is present only in small amounts in the liver during the first 20 days of postnatal life. During this period neither nuclear RNA synthesis nor induction of some liver enzymes can be affected by the glucocorticosteroids (Van der Meulen and Sekeris[31]); (d) *in vitro* transcription of nuclear fractions can be affected by cortisol or dexamethasone only in the presence of receptor G but not in that of A or B (Sekeris and Van der Meulen[96]).

Litwack and co-workers[97,98] have shown the presence of radioactive peaks (I–IV) after chromatography on DEAE-Sephadex of rat-liver cytosol derived from rats injected for 45 min with labelled cortisol in inducing doses. Peaks I and III bind anionic metabolites of cortisol whereas II and IV bind unmetabolized hormone. It seems that peak II has characteristics similar to binder G.

In thymus glucocorticosteroid receptors have been detected both in the cytosol and in the nucleus [99–101]. Contrary to the situation in the liver where most of the binding capacity is present in the cytosol, the thymus nuclei possess most of the cortisol or dexamethasone binding capacity. The nuclear receptor has a molecular weight of approximately 150000 in low ionic strength media and can be dissociated to 4-S molecules by salt[101]. It binds both natural and synthetic glucocorticosteroids with very high affinity and specificity. Evidence has been presented that this receptor is directly involved in the inhibitory action of the glucocorticoids on ribosomal RNA synthesis[34]. The cytosol appears to contain two binders, one similar to transcortin, as in the liver cytosol, and a 4-S receptor specific for glucocorticosteroids, natural and synthetic[102].

Similar receptor proteins have been detected in normal or malignant lymph cells[24,25,103–105], fibroblasts[106], lung tissues[107,108], lactating mammary glands[109], mammary tumours[110], and hepatoma cells[111,112]. A prominent feature of the cytosol receptor–hormone complex is its binding to chromatin or DNA[113,114]. As already mentioned for other hormones, the steroid receptor is activated by increasing ionic strength or temperature[115] and can then bind to nuclear material[116]. The nuclear-binding sites are destroyed by DNAase I (ref. 117).

(vi) Ecdysone-binding proteins

As the last steroid hormone to be considered, ecdysone should be mentioned. There are indications of ecdysone-binding proteins in the blood of insects (Emmerich[118]). They seem to fulfill a transport function similar to the well-studied serum transport protein in mammals. They have, however, not yet been isolated.

The relationship between ecdysone and the so-called "blood-factor" responsible for the development of spermatids (Kambysellis and Williams[119]) is not yet clear. Ecdysone is not active in this bioassay.

Recent experiments by Emmerich[120] have demonstrated the presence of ecdysone-binding proteins both in the cytosol and in the nucleus of target

TABLE II

Effects of hormones on RNA synthesis

Hormone	Target Tissue	Biological effect	Effect on RNA synthesis	Ref.
Oestrogens	uterus	growth	Increase in RNA synthesis; actinomycin and α-amanitin prevent hormone effect, increase in activity first of RNA polymerase B, then of A.	124–132
	chick oviduct	growth and differentiation	Increase in RNA synthesis, appearance of mRNA coding for ovalbumin.	149–151
Progesterone	uterus	differentiation	Increased RNA synthesis.	158a
	chick oviduct	growth and differentiation	Increased RNA synthesis, appearance of mRNA coding for avidin.	152–157
Androgens	prostate and seminal vesicle	growth and differentiation	Increased RNA synthesis, increased RNA-polymerase activity.	159–174
Aldosterone	bladder and kidney	Na retention	Increased incorporation of RNA precursors into RNA, increased activity of RNA polymerase by inhibitors of RNA synthesis.	175–183
Glucocortico-steroids	liver	gluconeogenesis	Increased incorporation of precursors into RNA, increased activity of RNA polymerase, increased template activity of chromatin, increased titre of mRNA coding for enzymes involved in gluconeogenesis.	185–189, 196, 197, 201, 202
	retina	differentiation	Accumulation of mRNA coding for glutamine synthetase.	203, 204
	thymus and other lymph cells	involution	inhibition of mainly ribosomal RNA synthesis	207, 208
Ecdysone	epidermis of insect larvae	sclerotization of cuticle	Increased RNA synthesis, inhibition of biological effects by inhibitors of RNA synthesis, increased formation of polysomes, increase of mRNA coding for DOPA-decarboxylase.	212–216, 219

cells. In general, these hormone-binding proteins are very similar to those described above for other steroids. However, they have not yet been amply characterized. In *Calliphora* larvae the presence of intracellular binding proteins has also been detected (Karlson *et al.*[121,122]).

Recently, Gorell *et al.*[123] have detected in the cytosol of the hepatopancreas of a Crustacean a protein binding a metabolite of α-ecdysone. According to these workers the metabolite may be involved in hormone action.

(c) Stimulation of RNA synthesis

As already indicated, the stimulation of RNA synthesis is one of the earliest effects of steroid hormones. This can be shown by several methods; some of the more customary experiments and their results are summarized in Table II, where some key references are given.

We will now consider in some detail the experimental results obtained with the various hormones.

(i) Oestrogens

The major target organ of oestradiol and other oestrogens is the uterus. Increase in RNA synthesis (and in protein synthesis, see later) after treatment with oestradiol was first observed by Mueller and co-workers (Mueller[124]; Ui and Mueller[125]). These effects are prevented by actinomycin D, also when applied locally in the uterus (Talwar and Segal[126]; Mueller[124]). According to Means and Hamilton[127], the first effects are seen within 2 min after administration of oestradiol *in vivo*.

Stimulation of RNA synthesis was measured in terms of the incorporation of precursors into RNA. Such *in vivo* experiments do not differentiate clearly between effects on precursor pool size, on membranes and on the overall activity of the RNA polymerase, respectively. This can be performed in an *in vitro* assay of nuclear fractions in the presence of a suitable mixture of nucleoside triphosphates and ions. Using this approach Gorski and co-workers[128,129] have shown an effect of oestrogens on RNA polymerase activity. Recently, Glasser *et al.*[130] have demonstrated an early action of oestrogens on polymerase B activity followed later by enhanced activity of polymerase A. As now established, polymerase B is responsible for the formation of the heterogeneous high molecular nuclear RNA, the RNA species containing mRNA sequences, whereas enzyme A is responsible for the synthesis of nucleolar RNA. Similar results were presented by Raynaud-

Jammet et al.[37,38], who showed in immature rat uterus an early α-amanitin-sensitive increase in RNA synthesis, followed by an increase in nucleolar RNA synthesis.

In accord with these findings are the results of Knowler and Smellie[131] showing a very early stimulation of heterogeneous RNA synthesis by oestradiol. Later on, ribosomal RNA is significantly enhanced; this is consistent with the fact that oestradiol generally stimulates growth of the target tissue[132].

Studies by Wittliff et al.[133] concerning oestrogen effects on RNA synthesis in amphibian liver show an early increase in rapidly labelled RNA which precedes an increase in protein synthesis, followed by the synthesis of RNA rich in guanine residues. The very early stimulation of these heterogeneous mRNA-containing giant molecules must be related to the specific processes induced by oestrogens.

A comparison of the time course of the induction of a specific protein like phosvitin in the liver of immature chicks with the synthesis of polysomal mRNA clearly showed such an oestrogen-dependent response (Jost et al.[134]).

Segal et al.[135] have reported that RNA isolated from uterus of oestradiol-treated rats can induce the same response as oestradiol when inoculated in the uteri of ovariectomized rats. This experiment, though fiercely criticized, has recently been confirmed by other independent groups[136–138].

Trachewski and Segal[139] have reported that RNA isolated after oestradiol treatment has a different base composition and nearest neighbour frequency. This is an indication of the specific response to oestrogens. Similar conclusions have been reached by Church and McCarthy[140] using the DNA/RNA hybridization technique.

The mechanism by which oestrogens increase RNA synthesis has been extensively studied. The following possibilities have been considered: (1) increase in the amount of the RNA polymerases; (2) increase in their activity; and (3) increase in the amount of template available for the enzyme.

The results obtained by Barker and Warren[141] and Teng and Hamilton[142], as well as those of Church and McCarthy[140] favour the third possibility, though other mechanisms cannot be ruled out. The experiments conducted by Barry and Gorski[143] suggest that oestradiol exerts an effect on the elongation of the synthesized RNA chains. Cox et al.[144] have demonstrated that there is indeed an oestrogen-induced increase in the template available for transcription by form B RNA polymerase. In addition, these authors 24 h after hormone stimulus measured an increase in extractable RNA polymerases A and B.

DNA synthesis is also stimulated by oestradiol but this effect is seen much later, *i.e.* after 48 h. This accords nicely with the well-known fact that one of the physiological actions of oestradiol is stimulation of cell proliferation.

Among the very early effects of oestradiol in the uterus are vasodilatation and an uptake of water from the circulation. This effect seems to be mediated by the liberation of histamine and serotonin, and can be seen shortly after oestradiol administration[145]. However, it is not clear whether this effect is due to an intracellular action on the part of oestradiol or to an action on the cell membrane. It will not be discussed here.

There are some indications that the water imbibition has something to do with cyclic AMP. It is, however, not clear whether cyclic AMP is released as a response to histamine and serotonin or as a response to oestrogens.

Effects on the synthesis of phospholipids[146] and of glycogen[147] have also been described. The biological implications of these findings are not yet clear. It should be mentioned that the vasodilatation is not inhibited by actinomycin D and is equally produced by oestriol; oestriol in turn has no effect on RNA metabolism.

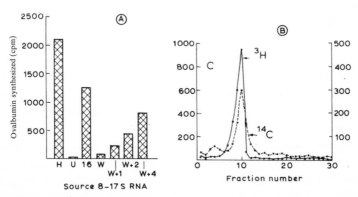

Fig. 3. Induction of mRNA by oestrogen in chick oviduct and *in vitro* translation of this mRNA into ovalbumin. (A), the effect of oestrogen on template activity of mRNA. Source of RNA: H, hen oviduct; U, unstimulated oviduct from 7-day old chicks; (16), the same after stimulation with 16 daily injections of diethylstilboestrol; (W) immature chicks received oestrogen for 16 days, then the hormone was withdrawn for additional 16 days; (W + 1, W + 2, and W + 4) W chicks to which oestrogen was readministered for 1, 2 and 4 days respectively. (B), experimental proof of ovalbumin as product of translation. mRNA from hens or treated chicks was translated in a cell-free rabbit reticulocyte lysate system, the protein formed precipitated with an antibody against ovalbumin, the precipitate solubilized and subjected to gel electrophoresis. Solid line: authentic ovalbumin labelled with ^{3}H; broken line: ^{14}C radioactivity from the reaction product. After Means *et al.*[150].

The effects of oestrogens on chick oviduct have been thoroughly examined by O'Malley and co-workers as well as by Schimke et al.[12,66,148,149]. A very early response is the stimulation of RNA polymerase activity after oestrogen administration. The newly synthesized RNA has been characterized by DNA/RNA hybridization experiments, suggesting the appearance of new RNA species (O'Malley[12]).

Using an in vitro protein-synthesizing system derived from reticulocytes, O'Malley and co-workers[66,150] succeeded in translating mRNA from oestrogen-primed oviducts into the specific oestrogen-induced protein ovalbumin (Fig. 3). Similar results have been obtained by Schimke and co-workers[151].

(ii) Progesterone

Progesterone is known to act on the oestradiol-treated uterus, initiating a so-called "secretory phase". In rabbit uterus progesterone induces the synthesis of a specific protein, uteroglobin. Studies of the influence of progesterone on RNA metabolism in this system have been started only recently. There are indications of stimulation (Bayer and Karlson, unpublished).

A much better system is the hen oviduct which, in response to progesterone, specifically synthesizes avidin[12]. The animal has to be treated beforehand with oestrogen to stimulate the oviduct. Stimulation of RNA synthesis is in this case readily demonstrated. After a temporary decrease in RNA synthesis (and RNA polymerase activity) 2 h after injection, the rate increases at 5–10 h and is maximal 24 h after injection[12]. The template activity of chromatin from induced oviducts is increased after progesterone treatment (O'Malley and McGuire[152]). In this case it has also been possible to show that progesterone treatment resulted in different RNA composition, as revealed by nearest neighbour analysis. Since this method is not a very refined one and gives only an overall picture, demonstrable differences borne out here may really mean specific action of the steroid hormone on RNA synthesis.

These findings were corroborated by hybridization studies of the newly synthesized RNA (O'Malley et al.[153–156]). Both with the basic hybridization technique and with the method of hybridization competition, it was shown that the RNA formed under the influence of progesterone contained new RNA species, indicating that new genes are transcribed under the action of progesterone.

Finally O'Malley and co-workers[157] have been able to demonstrate in an *in vitro* synthesizing system the synthesis of the specific protein (avidin) using an mRNA fraction derived from oviduct polysomes of progesterone-treated chicken. This RNA can be detected 6 h after progesterone administration and reaches a maximum 18–24 h after application of the hormone (Chan *et al.*[158]).

(iii) Androgens

The effect of androgens on precursor incorporation into RNA of target tissues such as the prostate and seminal vesicles has been well established.[159–163]. Nuclei isolated from target organs of animals subjected to androgen action also show an increased capacity for RNA synthesis[164]. Most of the RNA stimulated is considered ribosomal RNA on the basis of its GC content[165]. An increase in prostatic RNA polymerase activity was also described by Davies and Griffiths[166], who observed a stimulation of the nucleolar enzyme by 5 α-dihydrotestosterone–receptor complexes. The effect was most obvious when chromatin was used as template. This confirms the results of Mainwaring *et al.*[167] demonstrating a rapid and pronounced effect of androgenic hormones on Mg^{2+}-dependent nucleolar RNA polymerase activity. An increase in template activity was also shown by Couch and Anderson[168]. In their experiments, this effect was paralleled by a marked increase in the nonhistone protein content of prostate chromatin. The experiments by Liao and colleagues[169] showed no gross increase in template activity of chromatin. In experiments of a similar nature, however, Breuer and Florini[170,171] demonstrated an increase in template activity of chromatin from muscles of androgen-treated animals.

Increase in RNA synthesis has been reported for mouse kidney, which is also a target organ for androgens (Kochakian[172]).

Fujii and Villee[173,174] have reported that RNA from seminal vesicles of testosterone-treated immature rats has the same effect as androgens increasing the weight and protein content of seminal vesicles of immature rats. RNA from non-treated control animals failed to do so. This suggests that the action of testosterone is mediated by specific RNA. These experiments, interesting as they are, should be confirmed and alternative explanations should be ruled out.

(iv) Aldosterone

The effects of aldosterone on the RNA synthesis of target organs such as the toad bladder and rat kidney have been extensively studied in connection with the regulatory action of the steroid on sodium transport. A very rapid effect of aldosterone on the incorporation of uridine or other RNA precursors into RNA is demonstrated, prior to the effects of the hormone on Na^+ transport[175-180].

Within an hour after administration to adrenalectomized rats, nuclear RNA synthesis in the kidney is markedly enhanced[181,182]. Liew et al.[183] have shown a stimulatory effect of aldosterone on both Mn^{2+} and Mg^{2+}-activated nuclear RNA polymerase of kidney and heart muscle 1, 2 and 3 h after hormone administration, respectively. The effects of the hormone can be blocked by inhibitors of RNA[184] and protein synthesis (see below).

(v) Glucocorticosteroids

Feigelson et al.[185] first demonstrated that cortisone increases the RNA content of rat liver. The stimulatory effect of glucocorticoid on RNA synthesis was subsequently demonstrated by incorporation studies involving radioactively labelled RNA precursors injected into intact or adrenalectomized rats[186-188], or by measuring the RNA synthetic capacity of isolated subcellular fractions from hormone-treated animals[189,190]. Using both approaches, a very rapid stimulation of RNA synthesis was detected. Within 15–30 min RNA polymerase activity is significantly enhanced[190]. AU-rich RNA is first affected[191,192] and, at a later stage, GC-rich RNA[191]. This accords nicely with recent findings of an early stimulation by cortisol of extranucleolar, α-amanitin-sensitive RNA and then of nucleolar, α-amanitin-resistant RNA synthesis[193]. A specific increase in the synthesis of tRNA has been described by Thompson and co-workers[194,195]. After 1 h of dexamethasone treatment tRNA^{Phe} was newly synthesized, judging by the actinomycin D sensitivity of the effect. The mechanism of the increased RNA synthesis has been studied. Increased template activity of chromatin for RNA synthesis is one of the factors involved[196,197]. Hybridization experiments suggest the appearance of species of RNA which were absent or present in only small amounts[198,199] in the controls. After 4 h of cortisone treatment, Vorob'ev and Konstantinova[200] detected a marked increase in RNA, which could be hybridized to repetitive sequences.

Using a crude protein-synthesizing system derived from rat liver, Lang et al.[201] demonstrated that nuclear RNA fractions from livers of cortisol-induced rats stimulated the formation of tyrosine aminotransferase, measured by its enzymatic activity. RNA from control animals also gave rise to enzyme activity, but to a much smaller degree. Using more advanced techniques, Schütz *et al.*[202] have recently unequivocally demonstrated the formation *in vitro* of another inducible enzyme, tryptophan oxygenase, after the addition of mRNA fraction from liver polysomes of cortisol-induced rats to an ascites cell-free protein-synthesizing system.

Glucocorticosteroids inhibit RNA synthesis in the thymus, lymph cells and fibroblasts. In the thymus, this inhibition can be seen either after *in vivo* administration of the hormone[203] or after *in vitro* addition to isolated thymocytes[204] or thymus nuclei[205]. Ribosomal RNA is mainly impaired[206,207]. The effects on RNA synthesis can be abolished if the cells are treated with actinomycin D during the first 10 min of glucocorticosteroid action[208]. It has therefore been postulated that the first effect of the hormone is to stimulate mRNA coding for molecules taking part in subsequent degradative processes[209]. There is evidence of transfer of polymerase A from the nucleoli into the extranucleolar space[210]; this has also been observed in other cases where nucleolar function has been affected[211].

(vi) Ecdysone

As mentioned in the introduction, the induction by ecdysone[9] of specific puffs was the basic observation leading to the formulation of the hypothesis that hormones act through gene activation. It is therefore not surprising that ecdysone, like other steroid hormones, stimulates RNA synthesis in target tissues.

Within 1 h after ecdysone administration, nuclear RNA synthesis is significantly enhanced in the epidermis of developing blowfly larvae[212–214].

α–Ecdysone   β–Ecdysone

Within a few hours, mRNA has accumulated in the cytoplasm, resulting in the formation of polysomal structures (Marmaras and Sekeris[215]). RNA isolated from epidermis nuclei of hormone-induced, but not from control animals, is translated in a crude rat-liver protein-synthesizing system into the enzyme dopa-decarboxylase, as detected by its enzymatic activity (Sekeris and Lang[216]), suggesting the formation of dopa-decarboxylase under the influence of ecdysone. The conclusions reached in this experiment, however, may be differently interpreted. The experiments were recently repeated using the reconstituted protein-synthesizing system of Schreier and Staehelin[217] and a messenger RNA isolated by fractionation on oligo (dT) cellulose[218]. The results clearly demonstrate the synthesis of dopa-decarboxylase encoded by insect mRNA[219]. Three to four times more translatable dopa-decarboxylase mRNA is contained in the mRNA fractions from ecdysone-induced than in those from non-induced larvae[219].

Another target tissue for ecdysone, the wing tissue of Saturnid silkmoths, has received considerable attention.

Within 6 h after ecdysone administration significant stimulation of RNA synthesis was observed by Wyatt and Linzen[220]. No difference in the base composition of the newly synthesized RNA could be detected[221]. However, the formation of polysomes which actively synthesize proteins is induced, showing that mRNA formation was brought about by ecdysone[222]. The newly synthesized RNA showed stimulatory action on protein synthesis by an *in vitro* protein-synthesizing system from *E. coli*[223]. DNA-RNA hybridization studies suggested a selective effect of ecdysone on RNA synthesis, taking into consideration the limitations of the hybridization technique[224].

(d) Stimulation of protein synthesis

There are two general ways of demonstrating an increase in protein synthesis: (i) measuring the incorporation of labelled amino acids into protein; (ii) measuring a specific protein. The latter method is applicable either to specific proteins *that are under hormonal control* (*e.g.* ovalbumin and avidin in the chick, uteroglobin in the uterus) or to enzymes that can easily be detected and judged in terms of catalytic activity. Increased enzyme synthesis is called "enzyme induction". It is justified, however, to speak of "induced protein synthesis" in the case of specific proteins as well.

Some of the enzymes and proteins inducible by steroid hormones are listed in Table III; a more detailed account of the experimental work on

TABLE III

Enzymes and other proteins induced by steroid hormones

Hormone	Organ	Protein	Ref.
Oestrogens	uterus	I.P., K.I.P.,	225–227
		acidic protein	230
		(pI 4–5)	
	chick oviduct	ovalbumin	12
	chick liver	phosvitin	231–233
Progesterone	uterus	uteroglobin	235–238
	chick oviduct	avidin	12, 234
Androgens	mouse kidney	β-glucuronidase	241
Glucocorticosteroids	liver	tyrosine aminotransferase	245
		tryptophan oxygenase	246
		alanine transaminase	247
		and other enzymes of	248
		gluconeogenesis	
	embryonic retina	glutamine synthetase	249, 251
Aldosterone	kidney	protein involved in Na	242, 243
		transport	
Ecdysone	insect epidermis	DOPA-decarboxylase	213, 254,
			255

individual hormones is given below.

The mechanism of enzyme induction has been studied in detail in bacterial systems, *i.e.* in prokaryotes. In eukaryotes, the mechanism is less clear. In the following paragraphs, only experimental observations are reviewed, the theoretical aspects will be discussed later (p. 39).

(i) Oestrogens

Total protein synthesis is enhanced in the uterus by oestrogens[125]. Ribosomal preparations isolated from such induced uteri show increased capacity for protein synthesis[132]. This reflects the growth-promoting effect of oestrogens.

Evidence for the induction of a specific protein in the uterus 30–60 min after the administration of oestrogens has been presented by Notides and Gorski[225]. The protein has been named induced protein (I.P.). Its role is still unknown. Baulieu *et al.*[226,227] have also detected the rapid induction of a so-called "key intermediary protein" (K.I.P.), which seems to be identical to I.P. On the basis of α-amanitin experiments they postulate that this protein is an effector of ribosomal RNA synthesis. (The general idea that ribosomal RNA synthesis is under the direct or indirect control of transcription products of the extranucleolar chromatin has been discussed by Muramatsu et al.[228] and Sekeris and Schmid[211,229], see also p. 51.)

Another specific protein induced by oestradiol is an acidic protein (pI 4–5) complexed to the F3 histones of the uterus[230]. Its biosynthesis is stimulated within 15 min after oestradiol administration. Its function is not known.

Specific protein synthesis has been studied in the chick oviduct system by the groups of O'Malley and Schimke. 30 h after the administration of oestrogens to immature chicks, synthesis of ovalbumin is detected. At 15 days the level of ovalbumin is 300 times that of the base-line concentrations. This is due to the increased accumulation of mRNA for ovalbumin (see above) and the formation of polysomes synthesizing ovalbumin.

In the liver of chicks a highly phosphorylated protein, phosvitin, is induced by oestrogens[231–233].

(ii) Progesterone

In the oestrogen-primed oviduct progesterone induces the synthesis of the specific protein avidin (O'Malley[12]; O'Malley and Korenman[234]). Synthesis was first evident at 10 h after progesterone administration. As in the case of ovalbumin, the synthesis of avidin was attained in an *in vitro* protein-synthesizing system by the addition of mRNA preparations. The results suggest a *de novo* synthesis of avidin, due to the appearance of the mRNA[157].

In the uterus, progesterone induces the formation of uteroglobin, a protein whose function is not yet fully understood[235–238]. The mechanism of this induction could be similar to that of avidin in the chick oviduct.

(iii) Androgens

General protein synthesis is enhanced in those tissues in which the androgens promote growth, *i.e.* where they act as anabolic hormones[161]. This is reflected in an increase in polysome quantity and activity (Mainwaring and Wilce[239,240]). In mouse kidney, the activity of the enzyme β-glucuronidase is stimulated by testosterone (Frieden *et al.*[241]). The effect can be abolished by actinomycin D.

(iv) Aldosterone

The abolition of the effects of aldosterone on Na retention by inhibitors of protein synthesis such as puromycin or cycloheximide suggests that the specific protein named AIP (Fanestil[242]) mediates the effects of the hormone. The possible function of this protein has been discussed by Fanestil and Edelman[243]. It may serve as a carrier for Na, as a factor activating the transport ATPase or as a protein involved in the increased production of ATP.

Alteration of the protein-synthesizing capacity of rat-kidney cortex ribosomes has been reported by Trachewsky et al.[244].

(v) Glucocorticosteroids

Glucocorticoids induce a series of enzymes in mammalian liver, such as tyrosine transaminase, alanine transaminase, tryptophan oxygenase, PEP-carboxylase, involved in the gluconeogenic metabolic pathway[245-248].

As shown by studies with inhibitors of protein synthesis or by immuno-chemical methods, there is an increase in the amount of enzyme protein synthesized during induction[249,250]. Most workers agree that this is due to the stimulation of formation of mRNA coding for them, *i.e.* as a result of transcriptional control. Alternative explanations advanced on the basis of studies with hepatoma cells in culture and emphasizing post-transcriptional control of protein synthesis will be considered on p. 42.

In other tissues the glucocorticosteroids exert a catabolic effect. The amino acids generated in this process are subsequently used in the liver for gluconeogenesis. In the embryonic retina cortisol induces the *de novo* synthesis of glutamine synthetase[251]. In this case, the control seems to be exerted at the post-transcriptional level[252,253].

(vi) Ecdysone and protein synthesis

In the epidermis tissue of larvae, ecdysone induces the formation of the enzyme dopa-decarboxylase involved in the synthesis of the tanning agent N-acetyldopamine[213,254]. Inhibitors of protein synthesis prevent the appearance of enzyme activity, strongly suggesting the *de novo* synthesis of the decarboxylase[255].

Recently the increased incorporation of radioactively labelled amino acids into the enzyme molecule was detected[256], directly corroborating the conclusions reached on the basis of the use of metabolic inhibitors. General protein synthesis is not stimulated in the epidermis at this developmental stage.

In the wing epidermis of silkmoth pupae, ecdysone induces a general stimulation of protein synthesis[222]. Specific proteins synthesized under the influence of ecdysone in this tissue have not yet been found. A similar general increase in protein synthesis has also been demonstrated in the fat body of larvae from different diptera[222a].

3. Thyroid hormones

(a) General considerations

The uncoupling effect, postulated by Martius[257], of thyroxine on the respiratory chain and oxidative phosphorylation, was for a long time considered the main mode of action of the thyroid hormones thyroxine and 3,5,3'-triiodothyronine. Initial work on the mechanism of action of these hormones thus focused on mitochondrial events. However, it later became obvious that neither the swelling of mitochondria nor the uncoupling of the oxidative phosphorylation could constitute the main effect of these hormones. If this were the case then other uncoupling substances should mimic the thyroid effects, especially the induction of metamorphosis which is one of the main developmental processes controlled by thyroid hormones. But experiments showed that 2,4-dinitrophenol, which uncouples phosphorylation from the respiratory chain, is not able to initiate metamorphosis in Amphibia[258].

Since metamorphosis involves the synthesis of a variety of new proteins including the induction of several enzymes, research on the mechanism of action of thyroid hormone gradually shifted to phenomena like those described in the section on steroid hormones, *i.e.* the uptake of thyroid hormones by the cell, their transport within the cell and eventually into the nucleus, and regulation of gene activity in response to thyroid hormones.

Apart from these cytoplasmic and nuclear events it should be kept in mind that changes in mitochondrial metabolism still need to be explained. Since mitochondria possess transcriptional[259,260] and translational elements[261], the influence of thyroid hormones on RNA and protein synthesis within mitochondria was also investigated. Details are discussed below; the results may be interpreted in terms of a concerted control of nuclear, cytoplasmic and mitochondrial processes by thyroid hormones.

(b) Thyroid hormone-binding proteins

Investigations on the binding of thyroid hormones have been greatly influenced by the work performed in steroid hormone systems, although the first reports on cellular binding sites for thyroxine were published as early as 1958[262]. Tata then described the electrophoretic separation of a thyroxine-binding protein in rat skeletal muscle. Similar data were later presented by Manté-Bouscayrol[263] about thyroxine binding in rabbit skeletal muscle and brain. In these early investigations, it could be proven that the cytoplasmic thyroxine-binding proteins were different from serum thyroxine-binding protein.

The physiological significance of the hormone-binding capacity of the respective tissues was demonstrated by Tata[264] in the best investigated object of thyroid-hormone research, i.e. larval development of anurans. Without further analyzing the localization of bound hormones within the cell, he could clearly demonstrate a marked correlation between metamorphic competence and the capacity of whole tadpoles to bind ^{125}I-labelled triiodothyronine at higher temperatures. Initially the uptake was shown to be a function of the appearance of new cell populations. The subsequent increase in thyroxine-binding capacity kept pace with the overall accumulation of new proteins. At later stages binding is enhanced by higher temperatures.

There is considerable evidence that thyroxine and triiodothyronine are bound to different proteins in rat-liver cytosol[265]. These two binders may be separated by DEAE-cellulose chromatography and show, when compared to serum proteins, a lower affinity for thyroxine and an affinity for triiodothyronine which is approximately equal to the corresponding affinity of serum proteins. (It is well known that the affinity of serum proteins is much lower for triiodothyronine than for thyroxine[266].)

In contrast to these studies of differential binding of both thyroid hormones in rat-liver cytosol are the results of experiments concerning the nuclear binding of thyroid hormones in nuclei of GH_1 cells (a rat pituitary tumour cell line)[267]. There, the uptake of thyroxine could be competed by triiodothyronine; for both hormones 5000 binding sites per cell nucleus were determined. These high affinity/low capacity binding sites are restricted to the nucleus and were not detected in either mitochondria or cytosol. The authors thus emphasize that a possible cytosol receptor may be unstable until it is transferred to the nucleus. This nuclear binding of thyroid hormones already points toward a possible role played by these hormones in gene control. This is further evidenced by the fact that a nuclear triiodothyronine-binding protein complex was extracted from rat-liver chromatin. Its

References p. 54

TABLE IV

Thyroid hormone-binding proteins

Hormone	Tissue	Name	Localization	Mol. wt. or (S)	Separation	Characteristics	Ref.
Thyroxine	rat skeletal muscle	TBP	Cytosol	—	Electrophoresis	Thyroxine firmly bound as opposed to triiodothyronine.	262
Thyroxine	rabbit skeletal muscle rabbit brain	C-TBP	Cytosol Cytosol	— —	Electrophoresis Electrophoresis	Mobility similar to serum albumine in reverse flow paper, agar gel and starch column electrophoresis. Similar to β- and α-globulin on paper or cellulose column electrophoresis.	263
Triiodothyronine	whole Xenopus larvae	—	105000 Xg sediment	—	—	Strong correlation to metamorphic competence.	264
Thyroxine	rat liver	T4-binding protein	Cytosol	100000	DEAE-cellulose Sephadex G100	T4-BP and T3-BP are distinguishable by their elution profile from DEAE-cellulose.	265
Triiodothyronine	rat liver	T3-binding protein	Cytosol	100000	DEAE-cellulose Sephadex G100		265
Thyroxine/Triiodothyronine	GH$_1$ cells	—	Nuclei	—	Binding to intact cells and nuclei	K_d smaller for T3 than for T4. Both hormones compete for binding sites. Possible transfer from cytoplasm to nucleus.	267
Triiodothyronine	rat liver	—	Nuclei (chromatin)	60000 70000	Gel filtration	Binding affected by proteolytic enzymes, not by DNAase or RNAase.	268

apparent molecular weight is between 60000 and 70000 and the complex cannot be destroyed by RNAase or DNAase, which contrasts with its sensitivity to proteolytic enzymes[268].

It may be concluded from these results (Table IV) that the uptake of thyroid hormones by target cells is regulated by the binding of these hormones to proteins in cytoplasm and nucleus. The differential affinities of serum, cytoplasmic and nuclear-binding proteins respectively for thyroxine and triiodothyronine point toward the predominant role of triiodothyronine as compared to that of thyroxine.

(c) Stimulation of RNA synthesis

We mentioned that the effects of thyroid hormones on transcriptional and translational regulation have been studied mainly in amphibian metamorphosis, where this process is controlled by these hormones. Metamorphosis is the striking developmental event enabling the larvae to undergo the adaptational changes necessary for survival in the new environmental conditions of adult life. This involves both regressive events such as the involution of tail and gills as well as stimulation of growth, transformation of organs, and changes in metabolic pathways. The biochemical basis of these changes has been extensively studied by Cohen[269], Frieden[270], Weber[271], Tata[272] and their co-workers. Although most of the work has been done in amphibia, mammalian tissues and especially rat liver have also been investigated. The synthesis of new enzymes characteristic of adult amphibia (such as the enzymes of the urea cycle) as well as the increase in those proteins which are needed in larger quantity during terrestrial life depend on an increased formation of RNA.

Not only messenger RNA has to be newly synthesized; the translational machinery of the target tissue is reorganized as well, calling for the synthesis of large amounts of ribosomal RNA[273]. Fig. 4 and Table V summarize some of the effects of the thyroid hormones on target tissues.

The dependence of both regressive and anabolic phenomena on RNA and protein synthesis was demonstrated by the use of inhibitors for both synthetic processes[284]. Actinomycin D blocked, in vitro, the regression of isolated tail tips[275,277] and prevented the formation of carbamoylphosphate synthase[279] in tadpoles. Delayed administration of the inhibitor did not block the synthesis of this enzyme, which is characteristic of the adult animal.

Subsequently, RNA metabolism was studied in different target organs for thyroid hormones. In tadpole livers, 45 min after administration of triiodo-

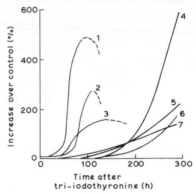

Fig. 4. Schematic representation of sequential stimulation of rate of RNA and phospholipid synthesis in relation to the increases in enzymes or protein synthesized on the precocious induction of metamorphosis in *Rana catesbeiana* tadpoles with tri-iodothyronine. Curve 1, rate of rapidly labelled nuclear RNA synthesis; curve 2, specific activity of RNA in cytoplasmic ribosomes; curve 3, rate of microsomal phospholipid synthesis; curve 4, carbamoyl-phosphate synthase activity; curve 5, cytochrome-oxidase activity/mg of mitochondrial protein; curve 6, appearance of serum albumin in the blood; curve 7, total liver protein/mg wet wt. The values are expressed as percentage increases over those in the non-induced control tadpoles. The broken lines in curves 1, 2 and 3 reflect the dilution of specific radioactivity in precursor molecules due to the onset of regression of tissues such as the tail and intestine. From Tata[286].

thyronine the specific activity of ^{32}P-labelled RNA gradually increases and, up to 48 h, different kinds of newly synthesized RNA appear[285]. The increase in specific activity of RNA is obvious even when no morphological changes can be detected[274]. The regression of tail tips *in vitro* is accompanied by an increase in the specific activity of their RNA, which reaches a maximum after four days of incubation[276,274].

The general pattern of RNA species in amphibia during metamorphosis was studied by Ryffel and Weber[276] by polyacrylamide gel electrophoresis. It revealed an overall increase in RNA synthesis showing different kinetics in different organs. In rat liver, the response of RNA synthesis to triiodo-thyronine showed a 30–40% increase even at 3–4 h after application of the hormone, and reached values of 100–200% after 11–16 h (ref. 281).

A detailed study of the newly synthesized RNA[283] by DNA/RNA-hybridization demonstrated the predominance of ribosomal RNA. This is compatible with the fact that a low dose of actinomycin D, which should only inhibit ribosomal RNA synthesis, abolishes all biological effects[286]. These results also agree with those showing, in isolated nuclei from rat-liver cells, a stimulation during the early phase of hormone action of the RNA polymerase responsible for ribosomal RNA synthesis[282].

TABLE V

Effect of thyroid hormones on RNA synthesis

Hormone	Target tissue	Biological effect	Effect on RNA synthesis	Ref.
Triiodothyronine	tadpole liver	changes in metabolic pathways	Increase in ^{32}P incorporation into RNA before any morphological change.	274
Thyroxine	different tissues of whole tadpoles	metamorphic reactions (regression, growth)	Sharp increase in nuclear ^{3}H-uridine incorporation, no extranuclear radioactivity up to 72 h (studied by autoradiography).	275
Thyroxine	tadpole brain, liver, hind limbs, tail muscle	regression and growth	Stimulation of all fractions of RNA separated by polyacrylamide-gel-electrophoresis. Tissue–specific difference in time course.	276
Triiodothyronine	isolated tadpole tails	tail regression	Stimulation of RNA synthesis parallel to protein synthesis.	277, 278
Thyroxine	tadpole liver	changes in metabolic pathways	Two peaks of RNA synthesis after 2 days and 12–15 days of hormone treatment.	279
Thyroxine	tadpole liver	changes in metabolic pathways	Rapidly labelled RNA in 6–10 S area has the properties of mRNA.	280
Triiodothyronine	rat liver	rise in activity of mitochondrial enzymes	Predominant synthesis of rRNA.	281, 282
Triiodothyronine	rat liver	rise in activity of mitochondrial enzymes	Preferential synthesis of rRNA, revealed by DNA/RNA hybridization.	283

It was also observed that in response to thyroid hormones the breakdown of pre-existing ribosomes is accelerated and followed by the formation of new polysomes[273]. This must involve not only the synthesis of new ribosomal RNA but also of messenger RNA.

A DNA-like base composition of newly synthesized rapidly labelled RNA was described by Nakagawa and Cohen[280] and by Blatt et al.[287]; this finding conflicts in some ways with the hybridization experiments performed by Wyatt and Tata[283]. It must be kept in mind, however, that the hybridization conditions employed by these authors do not favour detection of an increase in mRNA synthesis.

A direct study of RNA polymerase activity after 4 days of hormone treatment by Griswold et al.[288] again showed a marked stimulation of both RNA polymerase forms, suggesting the simultaneous stimulation of the synthesis of different RNA species. In rat liver, Smuckler and Tata[289] found no striking differences in the two main forms of RNA polymerase, but they did observe a stimulation of an additional, α-amanitin-insensitive peak.

Kim and Cohen[290] determined the template capacity for E. coli RNA polymerase using chromatin from thyroxine-treated tadpole livers. As for steroid hormones, it was shown that thyroid hormones also lead to a modification of the template activity of target-tissue chromatin. A stimulation exceeding that of untreated controls by 20–50% again points to the principle of hormonal regulation on the transcriptional level.

TABLE VI

Stimulation of protein synthesis by thyroid hormones during metamorphosis of amphibia

Hormone	Organ	Protein	Ref.
Thyroxine	tadpole liver	carbamoyl-phosphate synthase	293
Thyroxine	tadpole liver	glutamate dehydrogenase	294
Thyroxine	tadpole liver	cytochrome oxidase	277, 299
Triiodothyronine	serum	serum albumin	291, 292
Thyroxine	tadpole liver (frog liver)	adult haemoglobin	295, 296
Thyroxine	isolated tail tip	cathepsin	297
		acid phosphatase	298
		ribonuclease	298
		deoxyribonuclease	298
		β-glucuronidase	298

The hormones listed are those used by the authors in their respective experiments. For further data see also tables in refs. 269, 270, 271, 277, 297, 300, 303.

(d) Stimulation of protein synthesis

The regressive and anabolic processes during amphibian metamorphosis can be prevented not only by actinomycin D but also by inhibitors of protein synthesis, e.g. puromycin[277,278]. The predominant production of ribosomal RNA and changes in the structural organization of the endoplasmic reticulum during metamorphosis again demonstrate changes in the protein-synthesizing capacity of the tissue involved.

The proteins which are newly synthesized to permit survival under the environmental conditions of adult amphibian life include enzymes of the urea cycle and adult hemoglobin as well as proteins which are already present before metamorphosis but are needed in higher concentrations during terrestrial life (e.g. albumin[291,292]).

Table VI shows some of the proteins induced during metamorphosis. The most detailed analysis has been carried out on carbamoylphosphate synthase synthesis. Its de novo synthesis after thyroid-hormone application has been proven by immunoprecipitation[293] and by its non-appearance after the inhibition of RNA synthesis[279]. Also by immunoprecipitation, Balinsky et al.[294] demonstrated the de novo synthesis of glutamate dehydrogenase after induction with thyroxine.

(e) Thyroid hormones and mitochondria

The proteins described as newly synthesized during metamorphosis are partly mitochondrial enzymes. Thus the influence of thyroid hormones on mitochondrial metabolism is in part a consequence of actions on the transcriptional and translational level, resulting in an increased formation of enzymes and other proteins of mitochondria. Sokoloff et al.[301] even deduce from their results that early effects of thyroid hormones on protein synthesis require the presence of mitochondria.

It was already mentioned that mitochondria were for a long time considered the only site of action of thyroid hormones, after Martius[257] described the uncoupling effect of thyroxine on the respiratory chain and oxidative phosphorylation. However the calorigenic effect could not be linked up to the metamorphic events.

Tata and collaborators[302,302a] demonstrated an effect of thyroxine on de novo synthesis of respiratory units. These findings were corroborated by the results obtained by Kadenbach[303] showing an increase in the capacity of the respiratory system under hyperthyreotic conditions due to the induction of respiratory enzymes by thyroid hormones. A 2.5- to 3.5-fold increase in the total amount of skeletal muscle mitochondria was caused by thyroxine[302,304]. The stimulatory effect of thyroid hormones on the synthesis of nuclear RNA and of cytoplasmic protein respectively is paralleled by a stimulation of mitochondrial RNA polymerase activity[305]. DNA/RNA hybridization experiments by Gadaleta et al.[306] demonstrate that the RNA synthesized in mitochondria after treatment with triiodothyronine is qualitatively different from control preparations. A stimulation of amino acid incorporation into proteins by isolated mitochondria from hormone-treated animals was observed by Roodyn et al.[307].

(f) Conclusions

The results discussed above provide ample evidence that the intracellular fate and the mechanism of action of thyroid hormones is very similar to that of steroid hormones. Thyroid hormones, especially triiodothyronine, are bound to specific proteins (hormone receptors) and transported to the cell nucleus where they act as "signals" for transcription. They are bound to chromatin, the template activity of chromatin is modified, the synthesis of rRNA as well as mRNA is increased, and new proteins are synthesized. A special feature is the stimulation of mitochondrial transcription and of mitochondrial biogenesis.

4. Peptide hormones

In the preceding sections, the induction of de novo synthesis of enzymes and other proteins by hormones was described. Enzyme induction is an efficient but rather slow mode of metabolic regulation. For rapid adaptation to new metabolic situations, mechanisms permitting almost instantaneous activation of enzymes would be valuable.

This goal is achieved by hormone-mediated changes in the activity of membrane-bound adenylate cyclase and consequently of the intracellular level of cyclic AMP. A detailed review of this mechanism is beyond the scope of this Chapter; it should be mentioned, however, that virtually all peptide hormones use this "cAMP pathway" to exert their action on the target tissue. The rather strict separation of the two mechanisms might well be the consequence of a convergent evolution of hormonal systems.

However, besides this rapid control of enzyme activity via cAMP concentrations, some peptide hormones exert "slow" effects, resulting in the regulation of enzyme biosynthesis. This aspect of peptide hormone action will be discussed here, together with intracellular responses to those hormones that stimulate growth and differentiation in target cells.

(a) Somatotropin

When somatotropin is injected into hypophysectomized rats, the rate of RNA synthesis is accelerated[308-311]. This effect, however, cannot be completely inhibited by actinomycin D[309], suggesting that the action of this hormone is not exerted at the transcriptional level only.

It was already mentioned that Widnell and Tata[282] demonstrated a stimulation of RNA synthesis by thyroid hormones in isolated nuclei. In the same series of experiments these authors also demonstrated an increase in RNA polymerase activity after application of growth hormone. The effects of both hormones were shown to be additive. Likewise, the hybridizability of newly synthesized RNA, which increased after administration of thyroid hormone, was also enhanced by growth hormone[283]. In both cases, the synthesis of ribosomal RNA was predominant. Studies on isolated RNA polymerases in rat liver[289] showed again an increase in the additional peak of RNA polymerase activity, which could be eluted from DEAE-Sephadex at low ionic strength.

The changes in transcriptional activity are paralleled by effects on the amount and functional state of ribosomes[308,313]. Moreover, hypophysectomy leads to a decrease in the ability of ribosomes to react with polyuridylic acid[308]. It thus seems that the different hormonal status of hypophysectomized and hormone-treated animals respectively is reflected on both the level of RNA and on that of protein synthesis.

(b) Insulin

Among other effects not to be discussed here, insulin induces the biosynthesis of the enzyme glucokinase[312,314-316]. The synthesis of this enzyme could be prevented by inhibiting RNA synthesis with actinomycin D[317]. This is in agreement with the results obtained by Morgan and Bonner[318] who described an increase in template activity of liver chromatin after treating diabetic rats with insulin. In conformity with these findings, rat-liver nuclear RNA shows an increase in specific activity after administration of the hormone[319].

This stimulation of RNA synthesis, however, can be blocked by puromycin[320] and its regulation thus seems to be closely linked to protein synthesis. Indeed, protein synthesis is depressed in skeletal muscle of alloxan-diabetic rats, and the number of polysomes is diminished; these deficiencies can be corrected by insulin[321–326]. This response to insulin was also observed when RNA synthesis was blocked by actinomycin D. Again, as in the case of growth hormone, the functional state of ribosomes differs in normal and diabetic rats[322]. Normal levels of hormone are required for correct rates of peptide-chain initiation[327].

Experiments with inhibitors of RNA and protein synthesis and comparison with the mechanism of enzyme induction by steroid hormones further strengthened the view that insulin, in contrast to steroid hormones, controls enzyme induction mainly by acting at the translational level[327–333]. This effect, however, can also be achieved by addition of dibutyryl cyclic AMP to hepatoma cell cultures[329–331].

(c) Corticotropin

In the case of steroidogenesis, induced by corticotropin, hormonal effects again are mediated through cyclic AMP. However, this hormone also affects the transcriptional activity in its target gland.

In a continuous flow system, adrenals were exposed to ACTH and radio-active uridine[335,336]. An increase in specific radioactivity was observed in all RNA species, especially in 18S RNA.

Previous experiments by Garren and co-workers[337] demonstrated the key role of a protein with a high turnover rate acting on the translational level. Studies by Mostafapour and Tchen[338,339] on steroidogenesis in hypophy-sectomized rats led these authors to conclude that the synthesis of mRNA for this labile regulating protein might be the controlling factor in the regenerative effect of ACTH on adrenal glands in hypophysectomized rats.

(d) Thyrotropin and long-acting thyroid stimulator

Other gonadotropic hormones which are known to act via cyclic AMP are thyrotropin and long-acting thyroid stimulator (LATS). With isolated thyroid cells, El Khatib et al.[340] showed that the stimulation of protein synthesis by LATS is in part dependent upon the synthesis of new RNA. TSH, in this system, increases the synthesis of RNA and protein only when given together with cyclic AMP.

(e) Erythropoietin

Erythropoietin is a glycoprotein governing erythrocyte development[342]. The hormonal effect can be measured by the rate of uptake of ^{59}Fe or [^{14}C] glucosamine into bone marrow cells[343,344]. In contrast to the peptide hormones described above, which must rapidly regulate metabolic changes, erythropoietin is a slow-acting hormone. It affects the cyclic AMP level in rat fetal liver cells[341], but had long been known[344–346] to act on the transcription level by stimulating RNA synthesis.

RNA and DNA synthesis are stimulated, but DNA synthesis seems to depend on proteins synthesized after the early increase in transcription[347]. The early stimulation of RNA synthesis yields a 9-S RNA fraction[348] within 1 h. Inhibition of protein synthesis during this time does not abolish the effect of erythropoietin on RNA synthesis[348–350]. Thus the early effect does not seem to be mediated by a newly synthesized protein.

When isolated nuclei of erythropoietin-responsive cells are incubated with the hormone, no effect on RNA synthesis is observed. Therefore a cytoplasmic factor was postulated which might mediate the effect of the hormone on nuclear RNA synthesis. One such factor has been described[351]. It stimulates RNA synthesis when added to target-cell nuclei. Its discovery led to studies on erythropoietin receptor proteins on the surface of erythropoietin-responsive cells. Interaction of the hormone with such a receptor would then cause the bone marrow cell cytoplasmic factor to be generated or released to effect an increase in nuclear RNA synthesis. In analogy to work on insulin receptors, such a receptor protein on the surface of bone marrow cells has been described[352].

5. Theories on intracellular mechanisms of hormone action

(a) General considerations

First, it should be emphasized again that this review deals only with intra-cellular actions of hormones. Mechanisms involving interaction of hormones with the adenylate cyclase system located in the cell membrane are therefore beyond the scope of this article.

The preceding sections described experimental data on intracellular actions of hormones. By contrast, we shall now adopt a more theoretical approach inasmuch as we shall try to evaluate, in the light of established data, the various interpretations, hypotheses and theories that have been put forward during the last two decades.

Since most work has been done on steroid hormones, our discussion will be mainly concerned with the mechanism of action of steroid hormones. The data presented above permit three generalizations: (*i*) steroid hormones interact first with specific proteins called steroid receptors, (*ii*) one of the early effects is stimulation of RNA synthesis, (*iii*) next, synthesis of specific proteins is stimulated, *i.e.* proteins (often enzymes) are induced. The latter two processes are also known as transcription and translation, respectively. Thus, our task is to "explain" hormonal control of transcription and/or translation.

The general principles governing transcription and translation are well known and need not be repeated here. It should be pointed out, however, that insight into these principles has been gained mainly by work on bacterial systems, *i.e.*, transcription of the prokaryotic genome and translation of RNA on bacterial ribosomes. The mechanism of enzyme induction through regulation of transcription as explained by the well-known model of Jacob and Monod[353] also refers to bacterial systems.

Following the first observations that enzyme induction in mammalian systems and puff induction in insects are among the effects of hormones (see above), it was tempting to draw an analogy to enzyme induction in bacterial systems. The model devised by Jacob and Monod was rather indiscriminately applied to eukaryotes. However, work in the last decade has shown that in eukaryotes the situation is much more complicated because of the complexity of the cell nucleus as an organelle and the structural and functional features of mammalian chromosomes listed in Table VII (p. 46). It is safe to say that this complexity is not accidental as an evolutionary result, it must have advantages over the simpler bacterial systems. We must admit, however, that we do not yet know what this advantage is.

Since hormonal systems* are rather late "inventions" in the evolutionary chain, the complexity may even have something to do with hormone action, either as a prerequisite or as a consequence.

We shall now discuss the various hypotheses and theories concerning hormone action on the basis of the facts listed in Table VII (p. 46).

---

* By "hormonal system" we mean all that is needed for hormonal control: the hormone production in specialized tissues, "receptors" for the hormone, the machinery for hormone action, and biochemical routes for inactivation and excretion.

(b) Control of transcription through changes in inorganic ion concentration

In 1963, Kroeger[380] observed that explantation of salivary glands of *Chironomus thummi* into media high in K^+ resulted in changes in puffing pattern in the giant chromosomes. Among the puffs induced were those that may also be induced by ecdysone. Kroeger[381,382] then developed the hypothesis that ecdysone does not act directly at the level of the chromosomes but restricts the permeability of the cell membrane and/or the nuclear membrane to K^+ and Na^+ (possibly also to Ca^{2+} and Mg^{2+}) in such a way that the intranuclear concentration ratio K^+/Na^+ is raised. As a result of this change in ion concentration, puffs are induced, *i.e.*, specific genes are activated.

Kroeger's basic experiment—puff induction in explanted salivary glands by media of high ionic strength—has been confirmed by other workers. It has also been shown that puffs are able to react in narrow concentration ranges (for review, see Lezzi and Robert[383]). In details, the results are conflicting; Clever[384] found high salt concentrations (above 0.19 M) effective regardless of whether KCl or NaCl is used (but see also ref. 383). Berendes[385] working with *Drosophila* found a difference in puffing patterns resulting from either KCl or ecdysone (for reviews, see Clever[386], Ashburner[387], Lezzi and Gilbert[388], Beermann[389]).

It seems unnecessary to go into all details. The main question is whether changing ion concentrations is indeed the physiological way in which ecdysone induces puffs as visible signs of transcription control, or whether ion treatment just "mimics" ecdysone treatment to some extent, and ecdysone acts at the level of the chromosome*.

Indeed, there may be many reasons why high ion concentrations and/or imbalances may result in puffing and eventually in stimulation of RNA synthesis in salivary glands. It is well known that isolated chromatin when used as template for RNA synthesis *in vitro* shows much higher template activity when the ionic strength of the medium is raised[390]. Such an effect may also occur in explanted glands. Also relevant in this context may be the observation that the transformation of cytoplasmic hormone receptors into the "nuclear" form is highly dependent on the ionic millieu (see p. 10). It

---

* Recently, Kroeger (personal communication) has further developed his hypothesis in assuming a double action on the part of ecdysone (or ecdysone–receptor complex, respectively): a "general" effect at the level of the nucleus leading to increased RNA synthesis, and the control of ion flow at the cell membrane resulting in the activation of specific genes (puffs) *via* changes in ion balance (see also Lezzi and Gilbert[388]).

cannot be ruled out that such transformation when forced to occur *in vivo* in the absence of ecdysone would mimic its action*.

As to the site of ecdysone action, Claycomb *et al.*[392] have shown in autoradiographic studies that labelled ecdysone is concentrated in the nuclei and associated with the chromosomes of *Drosophila virilis*. Also, ecdysone-binding proteins have been extracted from nuclei and from chromatin of *Drosophila* salivary glands[120]. This is in conflict with the hypothesis that ecdysone acts on the cell membrane.

Another conflicting piece of evidence is the fact that ecdysone is able to act on isolated nuclei in stimulating RNA synthesis. This effect is apparently independent of ions; it can also occur in media free of Na^+ and K^+ (ref. 393). Ecdysone effects in this system are also ion-dependent, but in a different way than predicted by Kroeger's work.

With other steroid hormones, effects have been observed not only with isolated nuclei but with isolated chromatin as well. In this case, there is no membrane on which such a mechanism (control of ion permeability) could work. This makes it very difficult to apply this hypothesis to other hormones.

From a theoretical point of view, too, it is difficult to see how the specificity of gene stimulation might be explained by the ion theory. Though it has been shown that different gene loci react in a very narrow range of ion concentration, *i.e.* that there is a mechanism permitting the exact 'measurement' of ion concentrations, it is hard to believe that so many variations in ionic milieu are possible as to allow an exact and independent regulation of so many genes. Moreover, when several hormones act on the same target tissue, the situation becomes even more complicated. And finally, the mechanism by which hormones can adjust the ionic milieu within the cell in such a delicate way remains to be explained; a simple stimulation of, *e.g.*, a "sodium pump" or a "potassium channel" presumably would not result in exact regulation. In our opinion, it therefore seems very unlikely that transcription is controlled by this mechanism.

(c) Enzyme induction through translational or post-transcriptional control

It was already pointed out that translation is a process that may be regulated in higher organisms. Factors affecting the rate of translation have been

---

* Such a mechanism could also explain the appearance of puffs as a result of temperature shocks (see *e.g.* Beermann[389]; Berendes[391]), since it is known that the transformation of cytosol receptor into nuclear receptor can be brought about by high temperature.

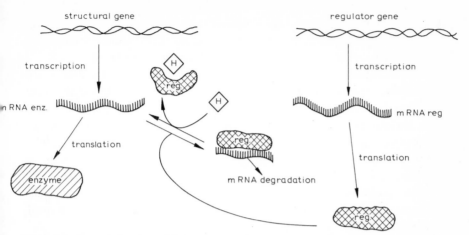

Fig. 5. Post-transcriptional control of enzyme synthesis according to Tomkins. Two genes are involved, the structural gene and a regulator gene. The regulatory protein (reg) is the product of transcription and translation of the regulator gene. The transcription product of the structural gene, mRNA enz., combines with a regulator protein and is degraded; little or no translation product (enzyme) is formed. When enzyme synthesis is induced by the hormone, the hormone is believed to combine with the regulatory protein, mRNA is no longer bound and is in turn released to the cytosol and translated.

described[379]. True control of translation should operate at the level of polysomes, either during initiation or elongation of the peptide chain, and should therefore be a cytoplasmic, not a nuclear event.

As far as the influence of steroid hormones is concerned, translational control has occasionally been postulated (Gorski[128]). However, since recently for most steroid hormones nuclear receptor proteins have been demonstrated, it is very likely that their site of action is the nucleus. We can therefore rule out "true" control of translation, *i.e.* a primary action of the hormone (or hormone–protein complex) at the level of the polysome.

Tomkins *et al.*[394–397] have postulated a mechanism of post-transcriptional control, outlined in Fig. 5. Two genes are involved, one structural, the other regulatory. The structural gene is transcribed into mRNA which in turn is exported to the cytosol and translated into the enzyme tyrosine trans-aminase. The regulatory gene is also believed to be transcribed and translated to yield a protein that interacts with the mRNA produced by the structural gene. This interaction is believed to inactivate the mRNA; it also renders the mRNA susceptible to degradation.

As a result of this inactivation and degradation of mRNA, very little protein (enzyme) is produced.

References p. 54

The inducer, *i.e.* the steroid hormone, is believed to interact with this protein*, resulting in its inactivation. Therefore more messenger can reach the cytoplasm undisturbed, and enzyme synthesis is enhanced.

The hypothesis of post-transcriptional control is based mainly on the observation that actinomycin when added 1–2 h after the steroid results in "superinduction", *i.e.*, a rate of enzyme synthesis exceeding that observed in the absence of actinomycin. Also, when the steroid is removed from the cell culture and the process of induction is reversed (de-induction), addition of actinomycin results in higher enzyme activity than is observed in its absence.

That actinomycin D leads to an increase in the concentration of specific proteins—and this is the basis for this theory—is frequently observed. Indeed, Tomkins *et al.*[397] published a table containing more than 50 references describing similar effects. It should be stressed, however, that the interpretation of these results is completely dependent on the assumption that actinomycin D acts exclusively on the level of transcription by blocking RNA synthesis. This assumption has been questioned[398,399]. Moreover, most experiments have been carried out with cultured minimal deviation hepatoma cells that do not necessarily reflect the situation *in vivo*.

In a detailed study of the rate of enzyme synthesis in the hepatoma cell system, Kenney *et al.*[400] obtained results conflicting with those described by Tomkins and his associates. They arrived at the conclusion that the inhibition of tyrosine aminotransferase degradation was the basis for the effect of actinomycin D on the enzyme concentration under the conditions of superinduction. In the same investigation, it was shown that cordycepin, which reduces the amount of mRNA connected to the polysomes[377,401], completely inhibits tyrosine aminotransferase induction by cortisol.

Another argument against the proposed model for superinduction was derived from experiments on the induction of ovalbumin synthesis by oestrogens. In this system, which also exhibits the superinduction phenomenon, Palmiter and Schimke[402] found no increase in the concentration of polysomal ovalbumin mRNA. In contrast to the Tomkins model for superinduction, these authors propose that different half-lives of mRNAs favour the relatively long-lived mRNAs for secretory proteins under the

---

* Tomkins calls this protein interacting with the messenger RNA "repressor". However, the term repressor is already defined by the Jacob and Monod model of regulation of transcription, and therefore it seems wise to avoid the use of the word "repressor" in the context of post-transcriptional control.

conditions of superinduction.

A more theoretical argument against the hypothesis of Tomkins *et al.* is that it implies a vicious circle of messenger production and degradation. In the case of simple metabolic chains and cycles, recent work on regulation of metabolism has demonstrated that vicious circles of this kind are generally avoided through various mechanisms (allosteric control of enzymes, inter-convertible enzymes, compartmentalization, etc.). It is hard to believe that such a vicious circle should operate in the cell nucleus and should function as the basis of enzyme induction in higher organisms. Though it is true that a large part of the primary transcription product (hnRNA) is degraded within the nucleus, very few investigators believe that the important mRNA sequences are constantly degraded before being used as template for protein synthesis.

Although, in our opinion, many experiments described in the foregoing sections favour the hypothesis of transcriptional control, the model of post-transcriptional control cannot yet be regarded as disproved. It should be pointed out, however, that the model introduces many more "unknowns" than it explains. For example, the function of the protein that combines with the messenger RNA (or pre-messenger RNA), rendering it at once untranslat-able and susceptible to degradation, is not understood. Moreover, the level of this protein in the nuclear sap must be carefully controlled to allow normal production of enzyme at a low level as well as induction by hormones to yield higher enzyme levels. It seems almost inescapable to postulate an additional control of transcription of the structural as well as the regulatory genes (coding for the regulatory protein) to guarantee this balance. If, how-ever, transcriptional control is assumed, it seems easier to think of hormonal influence on this transcriptional control rather than to accept the mechanism of post-transcriptional control outlined above.

(d) Hormonal control of transcription at the level of the chromosome

The main features of the organization of chromosomes of higher organisms which are different from bacterial chromosomes are listed in Table VII, together with some functional aspects of transcription and translation. On the basis of these facts, several models for the structural and functional organization of mammalian chromosomes have been suggested. Details of these models will be discussed elsewhere in this treatise; however, a short recapitulation of their salient features seems necessary for our discussion of

TÁBLE VII

Some properties in which the genome of higher organisms differs from that of prokaryotes

1. Structural organization
   (a) The amount of DNA per genome is up to 10000 times the DNA of bacteria[354].
   (b) Nuclear DNA is organized not in one, but in several chromosomes.
   (c) The total DNA is composed of sequences of different repetitiveness, ranging from highly reiterated simple sequences (e.g. satellite DNA) to unique genes[355].
   (d) The highly reiterated genes for ribosomal RNA are located in a morphologically distinct structure, the nucleolar organizer.
   (e) Symmetric sequences also occurring in regulatory regions of prokaryotes[356,357] facilitate the formation of hairpin-like structures in eukaryotic DNA and RNA[358,359].
   (f) Eukaryotic chromosomes contain at least as much permanently associated chromosomal proteins as DNA. Chromosomal proteins fall into two main classes, histones and "non-histone proteins". The latter is presumably a rather heterogeneous group. Some proteins presumably have a regulatory function.

2. Transcription
   (a) There are several DNA-dependent RNA polymerases with different functions present in higher organisms, adding another tool for gene regulation[360].
   (b) Factors affecting the transcription rate have been described[360-366].
   (c) RNA is synthesized as a high molecular weight precursor; this necessitates mechanisms for processing this giant "heterogeneous nuclear RNA (hnRNA)" into the mRNA finally translated[359,367-371].
   (d) The majority of the RNA synthesized does not leave the nucleus; it may or may not have a regulatory function[372].
   (e) Polyadenylate sequences are added to most mRNA species after transcription[373-378].
   (f) Methylation is another step in processing mRNA[378a].

3. Translation
   (a) For translation, the mRNA has to be transferred from the nucleus (site of transcription and processing) to the cytoplasm.
   (b) Ribosomes of eukaryotes are considerably larger and contain more protein species.
   (c) The ribosomes are in part attached to the endoplasmic reticulum.
   (d) Factors regulating initiation of translation, and apparently discriminating between different mRNA's, are known[379].

hormone action.

In this discussion we will use the term "structural gene" to designate those DNA sequences coding for a functional mRNA that is translated. Also the terms "operator", "promoter", and "repressor" are used in the same functional sense as in the classical Jacob and Monod theory.

Georgiev[403] proposed a model according to which a chromosome consists of transcriptional units (transcriptions) composed of a relatively large and partly or largely redundant acceptor zone where regulatory proteins can bind. At one end of the acceptor zone is the promoter site, at the other, a structural zone carrying one or more structural genes (Fig. 6a). When RNA polymerase binds to the promoter site, transcription can occur provided

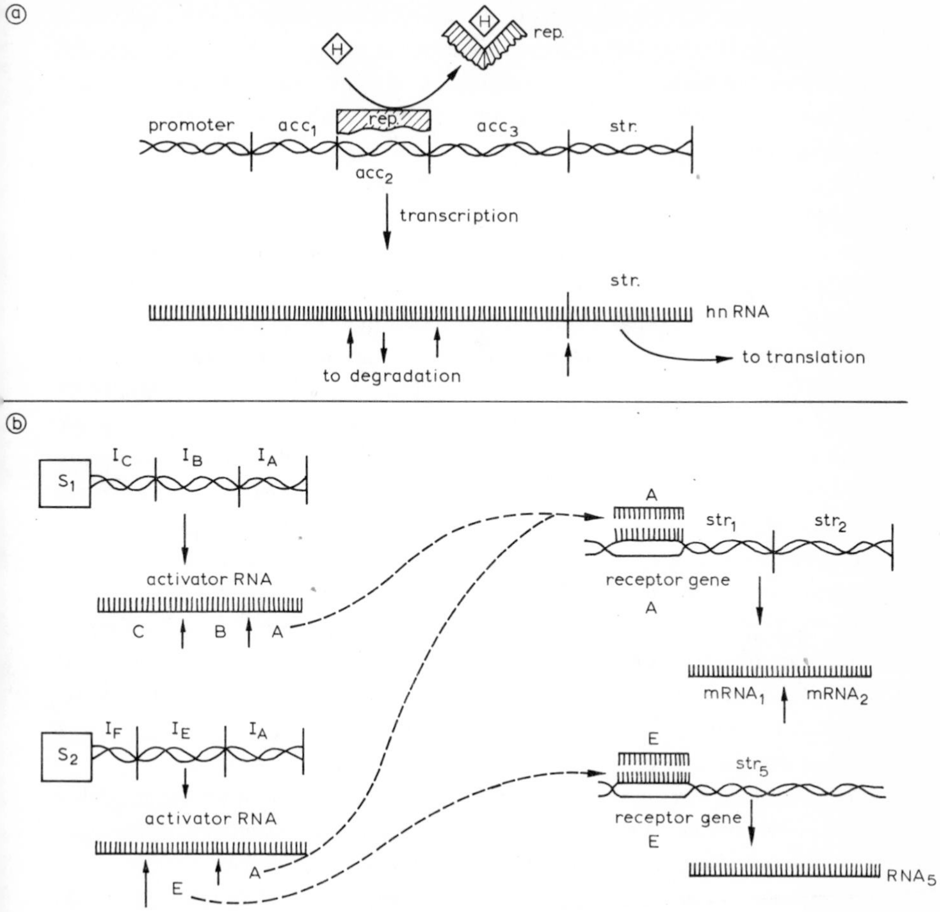

Fig. 6. (a) Scheme of the Georgiev model of chromosome organization and regulation of transcription. The DNA contains a promoter region (left), an acceptor zone (acc., center) and a structural gene (str, right). The acceptor zone contains several acceptor sites where regulatory proteins can bind and block transcription. If the whole acceptor region is free, transcription can occur and yield hnRNA containing the informative mRNA on the 3′ end. (b) Scheme of the Britten and Davidson theory of gene regulation. On the left are the sensor regions (S_1, S_2) attached to integrator genes. When the sensor receives a signal, the integrator genes are transcribed into activator RNA, which in turn diffuses to the receptor genes and initiates transcription of the corresponding structural genes (str_1, str_2, str_5). Two sets of integrator genes with two different sensors are shown; both contain the integrator gene I_A (a redundant gene), containing the information for activator RNA_A. Thus, activation of S_1 or S_2 will induce transcription of str_1 and str_2, while only activation of S_2 will induce transcription of str_5.

References p. 54

that the acceptor zone contains no blocking repressor proteins; removal of repressors may occur through combination with a hormone. Result of transcription is the high molecular weight nuclear RNA carrying only on its 3′ end an informative sequence. This hnRNA is then processed to yield functional mRNA[404,405].

A similar model has been proposed by Paul[406]; the main additional features are a proposal for an unwinding mechanism necessary to allow the DNA-dependent RNA polymerase to start transcription, and an explanation for the evolution of the highly reiterated sequences. Another difference is that the bulk of the reiterated sequences are meaningless; they are believed to be derived from nonsense mutation of an original structural gene amplified several times. These nonsense regions are transcribed but not translated; the complementary RNA sequences in hnRNA are destroyed during processing.

An even more sophisticated and elaborate model of chromosome structure has been proposed by Crick[407]. The model is mainly based on the structural features of giant chromosomes and takes into account the requirement for recognition sites in gene control. These sites are believed to be unwound stretches of the originally double helical DNA held in this position by proteins, possibly histones. They are believed to be located in the condensed bands that also contain the necessary operator and promoter sequences and may be highly redundant. The structural genes are believed to be in the interband regions. Altogether, one band plus one interband region are supposed to be one genetic complementation group.

It should be stressed that apart from this basic postulate, the model is more concerned with the structural organization of the chromosome than with functional regulation of transcription. However, Crick points out—and this is important in our context—that highly specific interactions are possible either between two proteins, between a protein molecule and an unpaired strand of nucleic acid, or between two unpaired nucleic acids that are complementary and can form a double helix (either DNA or DNA–RNA hybrid). Highly specific interaction (recognition) between a protein molecule and a paired stretch of nucleic acid is less likely for stereochemical reasons.

As far as regulation is concerned, all models discussed above are based on the assumption that this is in principle very similar to the situation in bacteria: Protein molecules equivalent to "repressors" are believed to bind to the regulatory DNA stretches and prevent transcription. Only removal of the repressor by the inducer (hormone or hormone–receptor complex) would allow transcription, a situation already depicted in Fig. 2a (p. 4). The main difference is that Crick proposes an unwound stretch of DNA as recognition site, while Georgiev and Paul do not.

Among the proteins present in the chromatin, the so-called "non-histone proteins" or "acidic proteins" are regarded as the most likely candidates for the role of regulatory proteins reacting with hormones. The histones are generally ruled out, mainly for the following reasons: (i) There are only few histone species present, at least not enough to allow a response to the many inducers presumably engaged in control; (ii) the amino acid sequence of histones shows very few changes during evolution; this is not compatible with the late appearance of hormonal systems; (iii) the rather small histone molecules may not allow tertiary structures intricate enough for the simultaneous recognition of a hormone molecule and a specific DNA sequence.

It should be mentioned that the specificity of histones might be altered and amplified through modification (acetylation, methylation, etc.); some authors have speculated along these lines. On the other hand, there are experimental data that hormone–receptor complexes interact with non-histone chromosomal proteins[75,76] and that these proteins are modified in hormone-treated cells[408]. It seems likely that these effects reflect the first step in induction of transcription, thus corroborating the postulate of non-histone proteins as gene regulators.

The models discussed above more or less attempt to combine the basic idea developed by Jacob and Monod with the complex situation of the eukaryotic chromosomes. They postulate a one-step mechanism of control. A more elaborate theory for gene regulation in higher cells has been put forward by Britten and Davidson[354]; their main postulate is a two-step mechanism of control within the nucleus. It is based upon the following considerations:

Simple external signals (like hormones) can induce changes in the rate of differentiation. These changes require the integrated activation of a large number of genes. Since the genome in higher organisms is much larger than the one in bacteria and contains repetitive sequences, the transcription of these repetitive sequences might be involved in the integrated regulation of gene activity. This idea is substantiated by experimental evidence; in the course of differentiation, repetitive sequences are transcribed in a cell-type specific distribution.

The model (cf. Fig. 6b) postulates sensor genes, where external agents bind in a sequence-specific way (these agents could be hormones or hormone–receptor complexes). The interaction of the external molecule with the sensor gene leads to the activation of an integrator gene, which is transcribed. The resulting activator RNA again forms a sequence-specific complex with a regulator gene, which is linked to the structural gene(s). The formation of a complex between the regulator gene and the activator RNA causes transcription of the structural gene(s). In conclusion, the interaction of an inducing agent with the sensor gene starts a sequence of events which finally results in the transcription of one or more structural gene(s).

The model includes the possibility of a gene redundancy on several levels. If the receptor genes are redundant, one signal can, through production of one activator RNA, induce the transcription of coding sequences for various proteins needed simultaneously. On the other hand, if integrator genes are redundant several signals will result in production of the same "activator RNA" (among others). Thus, redundancy in integrator as well as receptor genes may contribute to the repetitiveness of DNA, and this repetitiveness is functional, not accidental.

It should be stressed that according to this model a large part of the non-translatable nuclear RNA is used for the purpose of regulation, while in the models developed by Georgiev and Paul this RNA is just a by-product without specific function. Another new assumption is interaction between a regulatory RNA and the genetic material, DNA; this possibility was later also taken into consideration by Crick[407]. Moreover, this model offers the opportunity of integrating the activity of many different genes by one external signal, which is exactly the situation observed in the case of a hormone acting as developmental stimulus[409].

. Indeed, this model is particularly suited to account for the mechanism of hormone action. This is also mentioned by the authors in a footnote citing multiple references. Regarding the action of ecdysone, for example, it is tempting to speculate that the activation of one puff could represent the transcription of a set of integrator genes which would then lead to the activation of a whole battery of structural genes. The model also accounts for the vast amount of RNA synthesized after a hormonal stimulus, produced even in cases where only few species of mRNA are needed; but most of the newly formed RNA would not be informative but regulatory. This concept would also explain the high hybridization capacity of nuclear RNA after a hormonal stimulus, since regulatory RNA would be derived from redundant genes. It might also be mentioned that the RNA fraction produced under oestradiol stimulation can produce the same effect as oestradiol[135]. This is more easily conceivable if the RNA has a "signal" character than if it is just the sum of mRNA's.

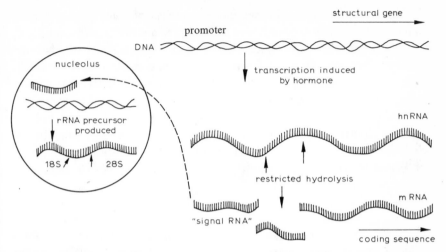

Fig. 7. Step-wise induction of messenger RNA and ribosomal RNA by a hormone. The hormone induces (through initiation of transcription or through derepression) the synthesis of heterogeneous nuclear RNA (hnRNA) which on restricted hydrolysis yields mRNA (used for translation) and a "signal RNA". The latter is transferred to the nucleolus and initiates the biosynthesis of the precursor of rRNA that is converted to rRNA.

The two-step mechanism postulated here also provides an explanation for simultaneous control of genes located on different chromosomes, as for example would be necessary during synthesis of proteins like hemoglobin, composed of two or more different polypeptide chains when their structural genes are located on different chromosomes. It could equally explain the correlation between the synthesis of messenger RNA and ribosomal RNA which has often been observed during the action of steroid hormones, e.g. glucocorticoids or androgens.

Work by Schmid and Sekeris[193] has shown that cortisol leads to a sequential stimulation of transcription. In the early phase the transcription of extranucleolar DNA by the α-amanitin-sensitive RNA polymerase B is stimulated; somewhat later, RNA synthesis catalyzed by polymerase A (insensitive to α-amanitin) is increased. Further experiments concerning the combined action of cortisol and α-amanitin applied later are compatible with a regulation mechanism such as outlined in Fig. 7. In this model, a somewhat different two-step mechanism is assumed. The "signal" cortisol induces the biosynthesis of hnRNA which during processing yields the messenger RNA for a protein as well as a "signal RNA" believed to be transferred to the nucleolus and there to stimulate rRNA synthesis. The possibility that signals are formed for the synthesis of other RNA species as well (tRNA?, a different mRNA or hnRNA?) is not excluded*.

The Britten–Davidson model leaves open the question whether the primary effect of hormone (or hormone–receptor complex) on the sensor is exerted by positive or negative control. Both are possible and would formally yield the same results. Also, the "activator RNA" could act either by positive control (stimulation of transcription, as assumed by the authors) or by negative control, i.e. derepression. Also, the nature and function of chromosomal proteins is not specified but the considerations given above (p. 49) on the role of non-histone proteins would apply here as well.

(e) Conclusions

The foregoing discussion has shown that it is as yet impossible to draw up a valid scheme representing the regulation of transcription in higher organisms. Several models have been proposed that explain many of the experimental data, but much more work will be needed before a final scheme of the process of transcription and its regulation will emerge.

Nevertheless, some general conclusions about the action of hormones can be drawn. First of all, the present authors feel that among the three theoretical possibilities discussed above, that of direct transcriptional control at the level of the chromosome is most likely. It cannot yet be decided if this control operates through a one-step or a two-step mechanism (the basic assumption made by Britten and Davidson). However, the initial event should be a recognition of a specific regulatory DNA sequence by the hormone–receptor complex; this may or may not be mediated by another chromosomal protein. As a result of this recognition, transcription by polymerase B is initiated, either by positive or by negative control as outlined in Fig. 2a and b. This interpretation is in line with Crick's postulate[407] about specific interaction between macromolecules. It should be added to Crick's considerations that a recognition of a small signal molecule like a hormone seems to be possible through a protein only, and not through a nucleic acid.

This theory is more or less a modification of the initial model developed by Jacob and Monod[353] for the regulation of gene activity by inducers. We believe that its principles, having proven to be useful in the regulation of gene

---

* Such a cascade mechanism might also explain the superinduction phenomenon used by Tomkins as the main argument for post-transcriptional control, provided some assumptions about differential inhibition by actinomycin of the various types of RNA synthesis are made.

activity in the prokaryotes, are most likely retained during evolution from prokaryotes to eukaryotes. The regulatory mechanism will presumably have evolved to greater complexity and sophistication without losing the basic principles.

It is quite conceivable that in the course of this evolution, the one-step mechanism changed to a two-step or even a multi-step mechanism. Cascade mechanisms of control are very common in biological systems; the control of phosphorylase activity by adrenalin or glucagon through cyclic-AMP and the various protein kinases, as well as the enzyme cascade of the blood-clotting system are just two examples. In such a system, positive and negative control can operate in series; even feedback control of essential steps seems possible.

Thus, the scheme for the mechanism of hormone action as outlined in Fig. 2 (p. 4) still seems to be the best representation of our present knowledge. It may be incomplete, omitting one or more steps in the control sequence, but it should not be entirely wrong. It may be mentioned that the salient features of the very first scheme given by Karlson[1,2] are retained in this more elaborate figure.

Assuming the basic idea that hormones are indeed the signal substances for control of gene activity, the experimenter is able to manipulate the system under investigation according to his needs, and the various experimental systems outlined in Sections 2–4 may provide very valuable tools for further elucidation of the regulation of gene activity in higher organisms.

REFERENCES

1 P. Karlson, *Dtsch. Med. Wochenschr.*, 86 (1961) 668.
2 P. Karlson, *Perspect. Biol. Med.*, 6 (1963) 203.
2a P. Karlson, *Naturwissenschaften*, 62 (1975) 126.
3 D. E. Green, *Adv. Enzymol.*, 1 (1941) 177.
4 O. Hechter, *Vitam. Horm. (Leipzig)*, 13 (1955) 293.
5 W. Dirscherl, 5. *Colloq. Ges. Physiol. Chem.*, (1955) 162.
6 C. Villee, D. D. Hagermann and B. B. Joel, *Recent Progr. Horm. Res.*, 16 (1960) 49.
7 P. Talalay and H. G. Williams-Ashman, *Recent Progr. Horm. Res.*, 16 (1960) 1.
8 N. Talal, G. M. Tomkins, J. F. Mushinski and K. L. Yielding, *J. Mol. Biol.*, 8 (1964) 46.
9 U. Clever and P. Karlson, *Exp. Cell Res.*, 20 (1960) 623.
10 C. Pelling, *Nature (London)*, 184 (1959) 655.
11 J. R. Tata, *Progr. Nucleic Acid Res. Mol. Biol.*, 5 (1966) 191.
12 B. O'Malley, *Trans. N.Y. Acad. Sci.*, 31 (1969) 478.
13 G. M. Tomkins and D. W. Martin Jr., *Annu. Rev. Genet.*, 4 (1970) 91.
14 G. Litwack (Ed.), *Biochemical Action of Hormones*, Academic Press, New York, 1972.
15 G. Raspe (Ed.), *Advances in Bioscience*, Vol. 7, Pergamon, Oxford, 1971.
16 P. P. Foa (Ed.), *The Action of Hormones*, Thomas, Springfield, Ill., 1971.
17 W. E. Stumpf, *Science*, 162 (1968) 1001.
18 S. Liao and S. Fang, *Recent Progr. Horm. Res.*, 27 (1969) 17.
19 J. H. Hutchinson and G. A. Porter, *Res. Commun. Pathol. Pharmacol.*, 1 (1970) 363.
20 A. Munck and T. Brinck-Johnsen, *J. Biol. Chem.*, 243 (1968) 5556.
21 E. V. Jensen, *Can. Cancer Conf.*, 6 (1965) 143.
22 D. Marver, J. Stewart, J. W. Funder, D. Feldman and I. S. Edelman, *Proc. Natl. Acad. Sci. (U.S.A.)*, 71 (1974) 1431.
23 W. Rosenau, J. D. Baxter, G. G. Rousseau and G. M. Tomkins, *Nature New Biol.*, 237 (1972) 20.
24 A. F. Kirkpatrick, R. J. Milholland and F. Rosen, *Nature New Biol.*, 232 (1971) 216.
25 N. Hollander and Y. W. Chiu, *Biochem. Biophys. Res. Commun.*, 25 (1966) 291.
26 A. W. Steggles and R. J. B. King, *Eur. J. Cancer*, 4 (1968) 395.
27 W. L. McGuire, G. C. Chamness and R. E. Shepherd, *Life Sci.*, 14 (1974) 19.
28 F. Bresciani, E. Nola, V. Sica and G. A. Puca, *Fed. Proc. Am. Soc. Exp. Biol.*, 32 (1973) 2126.
29 J. D. Baxter and G. M. Tomkins, *Proc. Natl. Acad. Sci. (U.S.A.)*, 65 (1970) 709.
30 M. Beato, M. Kalimi and P. Feigelson, *Biochem. Biophys. Res. Commun.*, 47 (1972) 1964.
31 N. van der Meulen and C. E. Sekeris, *FEBS Letters*, 33 (1973) 184.
32 J. H. Clark and J. Gorski, *Science*, 169 (1970) 76.
33 E. V. Jensen and E. R. DeSombre, in: G. Litwack (Ed.), *Biochemical Mechanisms of Hormone Action*, Academic Press, New York, 1972, p. 215.
34 N. van der Meulen, A. D. Abraham and C. E. Sekeris, *FEBS Letters*, 25 (1972) 116.
35 M. Beato, W. Braendle, D. Biesewig and C. E. Sekeris, *Biochim. Biophys. Acta*, 208 (1970) 125.
36 M. Arnaud, Y. Beziat, J. L. Borgna, J. C. Guilleux and M. Mousseron-Canet, *Biochim. Biophys. Acta*, 254 (1971) 241.
37 C. Raynaud-Jammet and E. E. Baulieu, *C. R. Acad. Sci. Ser. D*, 268 (1969) 3211.
38 C. Raynaud-Jammet, M. G. Catelli and E. E. Baulieu, *FEBS Letters*, 22 (1972) 93.
39 E. V. Jensen and H. Jacobson, *Recent Progr. Horm. Res.*, 18 (1962) 387.
40 E. V. Jensen, T. Suzuki, T. Kawashima, W. E. Stumpf, P. W. Jungblut and R. E. DeSombre, *Proc. Natl. Acad. Sci. (U.S.A.)*, 59 (1968) 632.

41 E. R. DeSombre, G. A. Puca and E. V. Jensen, *Proc. Natl. Acad. Sci. (U.S.A.)*, 64 (1969) 148.
42 J. Gorski, D. Toft, G. Shyamala, D. Smith and A. Notides, *Recent Progr. Horm. Res.*, 24 (1968) 45.
43 R. J. B. King and J. Gordon, *Endocrinology*, 34 (1966) 431.
44 E. J. Peck Jr., J. Burgner and J. H. Clark, *Biochemistry*, 12 (1973) 4596.
45 H. R. Maurer and G. R. Chalkley, *J. Mol. Biol.*, 27 (1967) 431.
46 G. P. Talwar, S. J. Segal, A. Evans and D. W. Davidson, *Proc. Natl. Acad. Sci. (U.S.A.)*, 52 (1964) 1059.
47 W. D. Noteboom and J. Gorski, *Arch. Biochem. Biophys.*, 111 (1965) 559.
48 T. Erdos, *Biochem. Biophys. Res. Commun.*, 32 (1968) 338.
49 H. Rochefort and E. E. Baulieu, *Endocrinology*, 84 (1969) 108.
50 H. Rochefort and E. E. Baulieu, *C. R. Acad. Sci. Ser. D*, 267 (1968) 662.
51 G. M. Stancel, K. M. T. Leung and J. Gorski, *Biochemistry*, 12 (1973) 2130.
52 A. C. Notides, D. E. Hamilton and J. H. Rudolph, *Endocrinology*, 93 (1973) 210.
53 E. V. Jensen, M. Numata, P. I. Brecher and E. R. DeSombre, in: R. M. S. Smellie (Ed.), The Biochemistry of Steroid Hormone Action, *Biochem. Soc. Symp.*, 32 (1971) 133.
54 P. W. Jungblut, S. McCann, L. Görlich, G. C. Rosenfeld and R. Wagner, in: C. Conti (Ed.), Research on Steroids, Vol. 4, Vieweg, Braunschweig, 1970, p. 213.
55 G. A. Puca, N. Nola, V. Sica and F. Bresciani, *Adv. Biosci.*, 7 (1971) 97.
56 R. J. B. King, V. Beard, J. Gordon, A. S. Pooley, J. A. Smith, A. W. Steggles and M. Vertes, *Adv. Biosci.*, 7 (1971) 21.
57 E. V. Jensen, *Acta Endocrinol. (Copenhagen)*, in press.
58 G. A. Puca, V. Sica and E. Nola, *Proc. Natl. Acad. Sci. (U.S.A.)*, 71 (1974) 979.
59 S. Mohla, E. R. DeSombre and E. V. Jensen, *Biochem. Biophys. Res. Commun.*, 46 (1972) 661.
60 E. R. DeSombre, S. Mohla and E. V. Jensen, *Biochem. Biophys. Res. Commun.*, 48 (1972) 1601.
61 J. André and H. Rochefort, *FEBS Letters*, 32 (1973) 330.
62 G. Shyamala-Harris, *Nature New Biol.*, 231 (1971) 246.
63 L. E. Clemens and L. J. Kleinsmith, *Nature New Biol.*, 237 (1972) 204.
64 K. R. Yamamoto and B. M. Alberts, *Proc. Natl. Acad. Sci. (U.S.A.)*, 69 (1972) 2105.
65 R. J. King and J. Gordon, *Nature (Lond.)*, 240 (1972) 185.
66 B. W. O'Malley and R. A. Means, *Science*, 183 (1974) 610.
67 E. E. Baulieu, A. Alberga, I. Jung, M. C. Lebeau, C. Mercier-Bodard, E. Milgrom, J. P. Raynaud, C. Raynaud-Jammet, H. Rochefort, H. Truong and P. Robel, *Recent Progr. Horm. Res.*, 27 (1971) 351.
68 E. Milgrom and E. E. Baulieu, *Endocrinology*, 87 (1970) 276.
69 E. Milgrom, M. Atger and E. E. Baulieu, *Steroids*, 16 (1970) 741.
70 L. E. Faber, M. C. Sandman and H. E. Stavely, *J. Biol. Chem.*, 247 (1972) 8000.
71 E. Milgrom, L. Thi, M. Atger and E. E. Baulieu, *J. Biol. Chem.*, 248 (1973) 6366.
72 M. R. Sherman, P. L. Corvol and B. W. O'Malley, *J. Biol. Chem.*, 245 (1970) 6085.
73 W. T. Schrader and B. W. O'Malley, *J. Biol. Chem.*, 247 (1972) 51.
74 A. W. Steggles, T. C. Spelsberg and B. W. O'Malley, *Biochem. Biophys. Res. Commun.*, 43 (1971) 20.
75 T. C. Spelsberg, A. W. Steggles, F. Chytil and B. W. O'Malley, *J. Biol. Chem.*, 247 (1972) 1368.
76 M. L. Chatkoff and J. A. Julian, *Biochem. Biophys. Res. Commun.*, 51 (1973) 1015.
77 S. Fang, K. M. Anderson and S. Liao, *J. Biol. Chem.*, 244 (1969) 6584.
78 S. Fang and S. Liao, *J. Biol. Chem.*, 246 (1971) 16.
79 W. I. P. Mainwaring, *J. Endocrinol.*, 45 (1969) 531.

80 W. I. P. Mainwaring and R. Irving, *Biochem. J.*, 134 (1973) 113.
81 J. A. Blaquier and R. S. Calandra, *Endocrinology*, 93 (1973) 51.
82 U. Gehring, G. M. Tomkins and S. Ohno, *Nature New Biol.*, 232 (1971) 106.
83 J. S. Norris, J. Gorski and P. O. Kohler, *Nature (Lond.)*, 248 (1974) 422.
84 F. S. French and E. M. Ritzen, *Endocrinology*, 93 (1973) 88.
85 I. Jung and E. E. Baulieu, *Nature New Biol.*, 237 (1972) 24.
86 G. Giannopolos, *J. Biol. Chem.*, 248 (1973) 1004.
87 G. A. Porter and I. S. Edelman, *Proc. Natl. Acad. Sci. (U.S.A.)*, 52 (1964) 1326.
88 I. S. Edelman and G. M. Fimognari, *Recent Progr. Horm. Res.*, 24 (1968) 1.
89 D. Ausiello and G. W. G. Sharp, *Endocrinology*, 82 (1968) 1163.
90 G. E. Swanneck, E. Highland and I. S. Edelman, *Nephron*, 6 (1969) 297.
91 G. W. Swanneck, L. H. Chu and I. S. Edelman, *J. Biol. Chem.*, 245 (1970) 5382.
92 M. Beato, W. Schmid and C. E. Sekeris, *Biochim. Biophys. Acta*, 263 (1972) 764.
93 M. Beato and P. Feigelson, *J. Biol. Chem.*, 247 (1972) 7890.
94 M. Koblinsky, M. Beato, M. Kalimi and P. Feigelson, *J. Biol. Chem.*, 247 (1972) 7897.
95 C. E. Sekeris and W. Schmid, *IVth Intern. Congr. Endocrin., Washington*, 1972, p. 408.
96 C. E. Sekeris and N. van der Meulen, *Acta Endocrin. (Copenhagen)*, in press.
97 K. S. Morey and G. Litwack, *Biochemistry*, 8 (1969) 4813.
98 S. Singer and G. Litwack, *Endocrinology*, 88 (1971) 1448.
99 B. P. Schaumburg, *Biochim. Biophys. Acta*, 261 (1972) 219.
100 C. Wira and A. Munck, *J. Biol. Chem.*, 245 (1970) 3436.
101 A. D. Abraham and C. E. Sekeris, *Biochim. Biophys. Acta*, 297 (1973) 142.
101a N. Kaiser, R. J. Milholland and F. Rosen, *J. Biol. Chem.*, 248 (1973) 478.
102 N. van der Meulen, K. Lipp and C. E. Sekeris, *Klin. Wochenschr.*, 52 (1974) 571.
103 A. F. Kirkpatrick, N. Kaiser, R. J. Milholland and F. Rosen, *J. Biol. Chem.*, 247 (1972) 70.
104 U. Gehring, B. Mohit and G. M. Tomkins, *Proc. Natl. Acad. Sci. (U.S.A.)*, 69 (1972) 3124.
105 M. Lippman, R. Halterman, S. Perry, B. Leventhal and E. B. Thompson, *Nature New Biol.*, 242 (1973) 157.
106 J. F. Hackney and W. B. Pratt, *Biochemistry*, 10 (1971) 3002.
107 G. Giannopulos, S. Mailay and S. Solomon, *Biochem. Biophys. Res. Commun.*, 47 (1971) 411.
108 D. Toft and F. Chytil, *Arch. Biochem. Biophys.*, 157 (1973) 464.
109 G. Shyamala, *Biochemistry*, 12 (1973) 3085.'
110 G. Shyamala, *J. Biol. Chem.*, 249 (1974) 2160.
111 J. D. Baxter and G. M. Tomkins, *Proc. Natl. Acad. Sci. (U.S.A.)*, 68 (1971) 932.
112 G. G. Rousseau, J. D. Baxter and G. M. Tomkins, *J. Mol. Biol.*, 67 (1972) 99.
113 M. Beato, M. Kalimi, M. Konstam and P. Feigelson, *Biochemistry*, 12 (1973) 3372.
114 J. D. Baxter, G. G. Rousseau, M. C. Benson, R. L. Garcea, J. Ito and G. M. Tomkins, *Proc. Natl. Acad. Sci. (U.S.A.)*, 69 (1972) 1892.
115 S. J. Higgins, G. G. Rousseau, J. D. Baxter and G. M. Tomkins, *J. Biol. Chem.*, 248 (1973) 5866.
116 S. J. Higgins, G. G. Rousseau, J. D. Baxter and G. M. Tomkins, *Proc. Natl. Acad. Sci. (U.S.A.)*, 70 (1973) 3415.
117 S. J. Higgins, G. G. Rousseau, J. D. Baxter and G. M. Tomkins, *J. Biol. Chem.*, 248 (1973) 5873.
118 H. Emmerich, *J. Insect Physiol.*, 16 (1970) 725.
119 M. Kambysellis and C. Williams, *Biol. Bull. (Woods Hole, Mass.)*, 141 (1971) 527.
120 H. Emmerich, *Gen. Comp. Endocrinol.*, 19 (1972) 543.
121 P. Karlson, C. E. Sekeris and H. R. Maurer, *Z. Physiol. Chem.*, 336 (1964) 100.
122 P. Karlson and G. Thamer, *Z. Naturforsch.*, 27b (1972) 1191.
123 T. A. Gorell, L. I. Gilbert and J. B. Siddall, *Proc. Natl. Acad. Sci. (U.S.A.)*, 69 (1972) 812.

124   G. C. Mueller, *Mechanism of Hormone Action*, Thieme, Stuttgart, 1965, p. 228.
125   H. Ui and G. C. Mueller, *Proc. Natl. Acad. Sci. (U.S.A.)*, 50 (1963) 256.
126   G. P. Talwar and S. Segal, *Proc. Natl. Acad. Sci. (U.S.A.)*, 50 (1964) 226.
127   A. R. Means and T. H. Hamilton, *Proc. Natl. Acad. Sci. (U.S.A.)*, 56 (1966) 1594.
128   J. Gorski, *J. Biol. Chem.*, 239 (1964) 889.
129   D. Noteboom and J. Gorski, *Proc. Natl. Acad. Sci. (U.S.A.)*, 50 (1963) 250.
130   S. R. Glasser, F. Chytil and T. C. Spelsberg, *Biochem. J.*, 30 (1972) 947.
131   J. T. Knowler and R. M. S. Smellie, *Biochem. J.*, 131 (1973) 689.
132   T. H. Hamilton, *Science*, 161 (1968) 649.
133   J. L. Witliff, K. L. Lee and F. T. Kenney, *Biochim. Biophys. Acta*, 269 (1972) 493.
134   J. P. Jost, R. Keller and C. Dierks-Ventling, *J. Biol. Chem.*, 248 (1973) 5262.
135   S. I. Segal, O. W. Davidson and K. Wada, *Proc. Natl. Acad. Sci. (U.S.A.)*, 51 (1965) 782.
136   M. M. Fencl and C. A. Villee, *Endocrinology*, 88 (1971) 279.
137   O. Unhjem, A. Attramadal and J. Sölna, *Acta Endocrinol. (Copenhagen)*, 58 (1968) 227.
138   P. O. Hubinont, F. Leroy and P. Galand (Eds.), *Basic Actions of Sex Steroids on Target Organs*, Karger, Basel, 1971, p. 112.
139   D. Trachewsky and S. J. Segal, *Eur. J. Biochem.*, 4 (1969) 279.
140   R. B. Church and B. J. McCarthy, *Biochim. Biophys. Acta*, 199 (1970) 103.
141   K. L. Barker and J. C. Warren, *Proc. Natl. Acad. Sci. (U.S.A.)*, 56 (1966) 1298.
142   C. S. Teng and T. H. Hamilton, *Proc. Natl. Acad. Sci. (U.S.A.)*, 60 (1968) 1410.
143   J. Barry and J. Gorski, *Biochemistry*, 10 (1971) 2384.
144   R. F. Cox, M. E. Haines and N. H. C. Carey, *Eur. J. Biochem.*, 32 (1973) 513.
145   C. M. Szego, *Fed. Proc. Fed. Am. Soc. Exp. Biol.*, 24 (1965) 1343.
146   Y. Aizawa and G. C. Mueller, *J. Biol. Chem.*, 236 (1961) 381.
147   J. Bitman, H. C. Cecil, M. L. Mench and T. R. Wren, *Endocrinology*, 76 (1965) 63.
148   T. Oka and R. T. Schimke, *J. Cell Biol.*, 43 (1969) 123.
149   P. O. Kohler, P. M. Grimley and B. W. O'Malley, *Science*, 160 (1968) 86.
150   A. R. Means, J. P. Comstock, G. C. Rosenfeld and B. W. O'Malley, *Proc. Natl. Acad. Sci. (U.S.A.)*, 69 (1972) 1146.
151   R. Palacios, D. Sulivan, N. M. Summers, M. L. Killy and R. T. Schimke, *J. Biol. Chem.*, 248 (1973) 540.
152   B. W. O'Malley and W. L. McGuire, *Endocrinology*, 84 (1969) 63.
153   B. W. O'Malley, W. L. McGuire, P. O. Kohler and S. G. Korenman, *Recent Progr. Horm. Res.*, 25 (1969) 105.
154   B. W. O'Malley, M. R. Sherman, D. O. Toft, T. C. Spelsberg, W. T. Schrader and A. W. Steggles, *Adv. Biosci.*, 7 (1971) 213.
155   B. W. O'Malley and W. L. McGuire, *Biochem. Biophys. Res. Commun.*, 32 (1968) 595.
156   B. W. O'Malley and W. L. McGuire, *Endocrinology*, 84 (1969) 63.
157   B. W. O'Malley, G. C. Rosenfeld, J. P. Comstock and A. R. Means, *Nature New Biol.*, 240 (1972) 45.
158   L. Chan, A. R. Means and B. W. O'Malley, *Proc. Natl. Acad. Sci. (U.S.A.)*, 70 (1973) 1870.
158a B. G. Miller, *Biochim. Biophys. Acta*, 299 (1973) 568.
159   D. L. Greenman, W. D. Wicks and F. T. Kenney, *J. Biol. Chem.*, 240 (1965) 4420.
160   S. Liao and S. Fang, *Vitam. Horm. (N.Y.)*, 27 (1969) 17.
161   C. D. Kochakian, in P. Karlson (Ed.), *Mechanisms of Hormone Action*, Thieme, Stuttgart, 1965, p. 192.
162   H. G. Williams-Ashman, *Cancer Res.*, 25 (1965) 1096.
163   F. R. Mangan, G. E. Neal and D. C. Williams, *Arch. Biochem. Biophys.*, 124 (1968) 27.
164   H. G. Williams-Ashman, R. C. Hancock, L. Jurkowitz and D. A. Silverman, *Recent Progr. Horm. Res.*, 20 (1964) 247.
165   S. Liao, R. W. Barton and A. H. Lin, *Proc. Natl. Acad. Sci. (U.S.A.)*, 55 (1966) 1593.

166   P. Davies and K. Griffiths, *Biochem. Biophys. Res. Commun.*, 53 (1973) 373.
167   W. I. P. Mainwaring, F. R. Mangan and B. M. Peterken, *Biochem. J.*, 123 (1971) 619–628.
168   R. M. Couch and K. M. Anderson, *Biochemistry*, 12 (1973) 3114.
169   S. Liao, A. H. Lin and R. W. Barton, *J. Biol. Chem.*, 241 (1966) 3896.
170   C. B. Breuer and J. R. Florini, *Biochemistry*, 4 (1965) 1544.
171   C. B. Breuer and J. R. Florini, *Biochemistry*, 5 (1966) 3857.
172   C. D. Kochakian, *Gen. Comp. Endocrin.*, 13 (1969) 146.
173   T. Fujii and C. A. Villee, *Proc. Natl. Acad. Sci. (U.S.A.)*, 57 (1967) 1468.
174   T. Fujii and C. A. Villee, *Proc. Natl. Acad. Sci. (U.S.A.)*, 62 (1969) 836.
175   I. S. Edelman, R. Bogoroch and G. A. Porter, *Proc. Natl. Acad. Sci. (U.S.A.)*, 50 (1963) 1169.
176   I. S. Edelman, G. A. Porter and R. Bogoroch, *Proc. Natl. Acad. Sci. (U.S.A.)*, 52 (1964) 1326.
177   J. Crabbe and P. DeWeer, *Nature (London)*, 202 (1964) 298.
178   T. R. Castles and H. E. Williamson, *Proc. Soc. Exp. Biol. Med.*, 119 (1965) 308.
179   G. J. Rousseau and J. Crabbe, *Biochim. Biophys. Acta*, 157 (1968) 25.
180   G. J. Rousseau and J. Crabbe, *Eur. J. Biochem.*, 25 (1972) 550.
181   L. Forte and E. J. Landon, *Biochim. Biophys. Acta*, 157 (1968) 303.
182   L. F. Congote and D. Trachewski, *Biochem. Biophys. Res. Commun.*, 46 (1972) 957.
183   C. E. Liew, D. K. Liu and A. G. Gornall, *Endocrinology*, 90 (1972) 488.
184   L. H. L. Chu and I. S. Edelman, *J. Membr. Biol.*, 10 (1972) 291.
185   M. Feigelson, F. Gros and P. Feigelson, *Biochim. Biophys. Acta*, 55 (1962) 495.
186   K. F. Jervell, *Acta Endocrinol. (Copenhagen), Suppl.*, 44 (1963) 88.
187   F. T. Kenney and F. J. Kull, *Proc. Natl. Acad. Sci. (U.S.A.)*, 50 (1963) 493.
188   N. Lang and C. E. Sekeris, *Life Sci.*, 3 (1964) 169.
189   O. Barnabei, B. Romano, G. di Bitonto, U. Tomasi and F. Sereni, *Arch. Biochem. Biophys.*, 113 (1966) 478.
190   N. Lang and C. E. Sekeris, *Life Sci.*, 3 (1964) 391.
191   F. L. Yu and P. Feigelson, *Biochem. Biophys. Res. Commun.*, 35 (1969) 499.
192   M. Beato, D. Doenecke, J. Homoki and C. E. Sekeris, *Experientia*, 26 (1970) 1074.
193   W. Schmid and C. E. Sekeris, *FEBS Letters*, 26 (1972) 109.
194   S. S. Yang, M. E. Lippman and E. B. Thompson, *Endocrinology*, 94 (1974) 254.
195   M. E. Lippman, S. S. Yang and E. B. Thompson, *Endocrinology*, 94 (1974) 262.
196   M. Dahmus and J. Bonner, *Proc. Natl. Acad. Sci. (U.S.A.)*, 54 (1965) 370.
197   M. Beato, J. Homoki, I. Lukacs and C. E. Sekeris, *Z. Physiol. Chem.*, 349 (1968) 1099.
198   J. Drews and G. Brawerman, *J. Biol. Chem.*, 242 (1967) 801.
199   D. Doenecke and C. E. Sekeris, *FEBS Letters*, 8 (1970) 61.
200   V. I. Vorob'ev and I. M. Konstantinova, *FEBS Letters*, 21 (1972) 169.
201   N. Lang, P. Herrlich and C. E. Sekeris, *Acta Endocrinol. (Copenhagen)*, 57 (1968) 33.
202   G. Schütz, M. Beato and P. Feigelson, *Proc. Natl. Acad. Sci. (U.S.A.)*, 70 (1973) 1218.
203   K. E. Fox and J. D. Gabourel, *Mol. Pharmacol.*, 3 (1967) 497.
204   M. H. Makman, S. Nakagawa and A. White, *Recent Progr. Horm. Res.*, 23 (1967) 195.
205   A. D. Abraham and C. E. Sekeris, *Biochim. Biophys. Acta*, 247 (1971) 562.
206   J. Drews, *Eur. J. Biochem.*, 7 (1969) 200.
207   J. Drews and L. Wagner, *Eur. J. Biochem.*, 13 (1970) 231.
208   K. M. Moshner, D. A. Young and A. Munck, *J. Biol. Chem.*, 246 (1971) 654.
209   A. Munck, *Perspect. Biol. Med.*, 14 (1971) 265.
210   N. van der Meulen, R. Marx, C. E. Sekeris and A. D. Abraham, *Exp. Cell Res.*, 74 (1972) 606.
211   W. Schmid and C. E. Sekeris, *Biochim. Biophys. Acta*, 312 (1973) 549.
212   P. Karlson and G. Peters, *Gen. Comp. Endocrinol.*, 5 (1965) 257.

213   C. E. Sekeris in: P. Karlson (Ed.), Mechanisms of Hormone Action, Academic Press, New York, 1965, pp. 149–167.
214   C. E. Sekeris and N. Lang, Z. Physiol. Chem., 341 (1965) 36.
215   V. J. Marmaras and C. E. Sekeris, Exp. Cell Res., 75 (1972) 143.
216   C. E. Sekeris and N. Lang, Life Sci., 3 (1964) 625.
217   M. Schreier and Th. Staehelin, J. Mol. Biol., 80 (1973) 329.
218   M. Hirsch and S. Penman, J. Mol. Biol., 80 (1973) 391.
219   E. Fragoulis and C. E. Sekeris, in preparation.
220   G. R. Wyatt and B. Linzen, Biochim. Biophys. Acta, 103 (1965) 558.
221   S. W. Applebaum, R. P. Ebstein and G. R. Wyatt, J. Mol. Biol., 21 (1966) 29.
222   G. R. Wyatt, in: G. Litwack (Ed.), Biochemical Actions of Hormones, Academic Press, New York, 1972, p. 385.
222a R. Arking and E. Shaaya, J. Insect Physiol., 15 (1969) 287.
223   A. J. Howells and G. R. Wyatt, Biochim. Biophys. Acta, 174 (1969) 68.
224   K. U. Lucas and G. R. Wyatt, J. Insect Physiol., 17 (1971) 2301.
225   A. C. Notides and J. Gorski, Proc. Natl. Acad. Sci. (U.S.A.), 56 (1966) 230.
226   C. R. Wira and E. E. Baulieu, C. R. Acad. Sci., 273 (1971) 218.
227   E. E. Baulieu, A. Alberga, C. Raynaud-Jammet and C. R. Wira, Nature (London), New Biol., 236 (1972) 236.
228   M. Muramatsu, N. Shimida and T. Higashinakawa, J. Mol. Biol., 53 (1970) 91.
229   C. E. Sekeris and W. Schmid, in: J. K. Pollak and J. W. Lee (Eds.), The Biochemistry of Gene Expression in Higher Organisms, Australia and New Zealand Book Company, 1973, p. 225.
230   K. L. Barker, Biochemistry, 10 (1971) 284.
231   P. J. Heald and P. M. Mc. Halan, Biochem. J., 92 (1964) 51.
232   O. A. Schjeider and M. R. Urist, Science, 124 (1956) 1242.
233   B. L. Jailkhani and G. P. Talwar, Nature (London), New Biol., 236 (1972) 239.
234   B. W. O'Malley and S. G. Korenman, Life Sci., 6 (1967) 1953.
235   H. G. Schwick, Wilhelm Roux' Arch. Entwicklungsmech. Org., 155 (1965) 283.
236   H. M. Beier, Biochim. Biophys. Acta, 160 (1968) 289.
237   A. T. Arthur and J. C. Daniel Jr., Fertil. Steril., 23 (1972) 115.
238   M. Beato and R. Baier, Biochim. Biophys. Acta, in the press.
239   W. I. P. Mainwaring and P. A. Wilce, Biochem. J., 130 (1972) 189.
240   W. I. P. Mainwaring and P. A. Wilce, Biochem. J., 134 (1973) 795.
241   E. H. Frieden, A. A. Harper, F. Chin and W. H. Fishman, Steroids, 4 (1964) 777.
242   D. D. Fanestil, Annu. Rev. Med., 20 (1969) 223.
243   I. S. Edelman and D. D. Fanestil, in: G. Litwack (Ed.), Biochemical Actions of Hormones, Academic Press, New York, 1970, p. 321.
244   D. Trachewsky, A. P. Nandi Majumbar and L. F. Congote, Eur. J. Biochem., 26 (1972) 543.
245   F. T. Kenney and R. M. Flora, J. Biol. Chem., 236 (1961) 2699.
246   H. L. Segal and N. S. Kim, Proc. Natl. Acad. Sci. (U.S.A.), 50 (1963) 912.
247   W. E. Knox, Br. J. Exp. Pathol., 32 (1951) 462.
248   E. Shrago, H. A. Lardy, R. C. Nordlie and D. O. Foster, J. Biol. Chem., 238 (1963) 3188.
249   F. T. Kenney, J. Biol. Chem., 237 (1962) 3495.
250   P. Feigelson and O. Greengard, J. Biol. Chem., 237 (1962) 3714.
251   A. A. Moscona, M. Moscona and N. Saenz, Proc. Natl. Acad. Sci. (U.S.A.), 61 (1968) 160.
252   R. Schwartz, Nature (London), New Biol., 237 (1972) 121.
253   M. Moscona, N. Frenkel and A. A. Moscona, Dev. Biol., 28 (1972) 229.
254   P. Karlson and C. E. Sekeris, Biochim. Biophys. Acta, 63 (1962) 489.
255   C. E. Sekeris and P. Karlson, Arch. Biochem. Biophys., 105 (1964) 483.
256   E. Fragoulis and C. E. Sekeris, in preparation.

257 C. Martius, in: Hormone und ihre Wirkungsweise, 5. *Colloq. Ges. Physiol. Chem.*, Springer, Berlin, 1954, p. 143.
258 F. Heinemann and R. Weber, *Helv. Physiol. Pharmacol. Acta,* 24 (1966) 124.
259 G. Attardi, Y. Aloni, B. Attardi, D. Ojala, L. Pica-Mantoccia, D. L. Robberson and B. Storrie, *Cold Spring Harbor Symp. Quant. Biol.,* 35 (1970) 599.
260 H. Küntzel, Z. Barath, I. Ali, J. Kind, H. H. Althaus and H. C. Blossey, in: Regulation of Transcription and Translation in Eukaryotes, *24. Colloq. Ges. Biol. Chem.,* (1973) 195.
261 P. Borst and L. A. Grivell, *FEBS Letters,* 13 (1971) 73.
262 J. R. Tata, *Biochim. Biophys. Acta,* 28 (1958) 91.
263 S. Manté-Bouscayrol, G. Cartonzon, R. Depieds and S. Lissitzky, *Gen. Comp. Endocrinol.,* 2 (1962) 193.
264 J. R. Tata, *Nature (London),* 227 (1970) 686.
265 S. Hamada, K. Torizuka, T. Miyake and M. Fukase, *Biochim. Biophys. Acta,* 201 (1970) 479.
266 J. R. Tata, *Recent Progr. Horm. Res.,* 18 (1962) 221.
267 H. H. Samuels and J. S. Tsai, *Proc. Natl. Acad. Sci. (U.S.A.),* 70 (1973) 3488.
268 M. I. Surks, D. Koerner, W. Dillmann and J. Oppenheimer, *J. Biol. Chem.,* 248 (1973) 7066.
269 P. P. Cohen, *Science,* 168 (1970) 533.
270 E. Frieden, *Recent Progr. Horm. Res.,* 23 (1967) 139.
271 R. Weber, in: *Biochemistry of Animal Development,* Vol. 2, Academic Press, New York, 1967, p. 227.
272 J. R. Tata, in: Wolstenholme and Knight (Eds.), *Control Processes in Multicellular Organisms,* 1970, p. 131.
273 J. R. Tata, *Nature (London),* 207 (1965) 378.
274 F. J. Finamore and E. Frieden, *J. Biol. Chem.,* 235 (1960) 1751.
275 H. Stadler, *Z. Zellforsch. Mikrosk. Anat.,* 117 (1971) 118.
276 G. Ryffel and R. Weber, *Rev. Suisse Zool.,* 78 (1971) 639.
277 J. R. Tata, in: C. Gual (Ed.), *Proc. 4th Int. Congr. Endocrinol.,* Mexico, 1968, p. 61.
278 J. R. Tata, *Dev. Biol.,* 13 (1966) 77.
279 H. Nakagawa, K. H. Kim and P. P. Cohen, *J. Biol. Chem.,* 242 (1967) 635.
280 H. Nakagawa and P. P. Cohen, *J. Biol. Chem.,* 242 (1967) 642.
281 J. R. Tata and C. C. Widnell, *Biochem. J.,* 98 (1966) 604.
282 C. C. Widnell and J. R. Tata, *Biochem. J.,* 98 (1966) 621.
283 G. R. Wyatt and J. R. Tata, *Biochem. J.,* 109 (1968) 253.
284 R. Weber, *Experientia,* 21 (1965) 665.
285 J. E. Eaton and E. Frieden, *Gumma Symp. Endocrinol.,* 5 (1968) 43.
286 J. R. Tata, in: *Wirkungsmechanismen der Hormone,* 18. Colloq. Ges. Physiol. Chem., Springer, Berlin, 1967, p. 87.
287 L. M. Blatt, K. H. Kim and P. P. Cohen, *J. Biol. Chem.,* 244 (1969) 4801.
288 M. D. Griswold and P. P. Cohen, *J. Biol. Chem.,* 247 (1972) 353.
289 E. A. Smuckler and J. R. Tata, *Nature (London),* 234 (1971) 37.
290 K. H. Kim and P. P. Cohen, *Proc. Natl. Acad. Sci. (U.S.A.),* 55 (1966) 1251.
291 A. E. Herner and E. Frieden, *J. Biol. Chem.,* 235 (1960) 2845.
292 T. P. Bennett and E. Frieden, in: M. Florkin and H. S. Mason (Eds.), *Comparative Biochemistry,* Academic Press, New York, 1962, p. 483.
293 R. L. Metzenberg, M. Marshall, W. K. Paik and P. P. Cohen, *J. Biol. Chem.,* 236 (1961) 162.
294 J. B. Balinsky, G. E. Shambaugh and P. P. Cohen, *J. Biol. Chem.,* 245 (1970) 128.
295 G. M. Maniatis and V. M. Ingram, *J. Cell Biol.,* 49 (1971) 372.
296 G. M. Maniatis and V. M. Ingram, *J. Cell Biol.,* 49 (1971) 380.
297 R. Weber, *Bull. Schweiz. Akad. Med. Wiss.,* 22 (1966) 27.
298 Y. Eeckhout, *Thèse Louvain,* Fac. Sci., 1965 (cit. in Ref. 297).

299   J. R. Tata, L. Ernster, O. Lindberg, E. Arrhenius, S. Pedersen and R. Hedman, *Biochem. J.*, 86 (1963) 408.
300   J. C. Perriard, *Wilhelm Roux' Arch. Entwicklungsmech. Org.*, 168 (1971) 39.
301   L. Sokoloff, P. A. Roberts, M. M. Januska and J. E. Kline, *Proc. Natl. Acad. Sci. (U.S.A.)*, 60 (1968) 652.
302   R. Gustafsson, J. R. Tata, O. Lindberg and L. Ernster, *J. Cell Biol.*, 26 (1965) 555.
302a O. Lindberg, *Naturwissenschaften*, 52 (1965) 379.
303   B. Kadenbach, *Biochem. Z.*, 344 (1966) 49.
304   N. J. Gross, *J. Cell Biol.*, 48 (1971) 29.
305   K. Schimmelpfennig, M. Rautenberg and D. Neubert, *FEBS Letters*, 10 (1970) 269.
306   M. N. Gadaleta, A. Barletta, M. Caldarazzo, T. de Leo and C. Saccone, *Eur. J. Biochem.*, 30 (1972) 376.
307   D. B. Roodyn, K. B. Freeman and J. R. Tata, *Biochem. J.*, 94 (1965) 628.
308   A. Korner and J. M. Gumbley, *Nature (London)*, 209 (1966) 505.
309   A. Korner, *Recent Progr. Horm. Res.*, 21 (1965) 205.
310   A. Korner, in: Th. Bücher and M. Sies (Eds.), Inhibitors, Tools in Cell Research, *20. Colloq. Ges. Biol. Chem.*, Springer, Berlin, 1969, p. 126.
311   A. Korner, in: G. E. W. Wolstenholme and J. Knight (Eds.), *Control Processes in Multicellular Organisms*, Churchill, London, 1970, p. 81.
312   K. L. Manchester, in: E. E. Bittar and N. Bittar (Eds.), *The Biological Basis of Medicine*, Academic Press, New York, 1968, p. 221.
313   D. C. N. Earl and A. Korner, *Arch. Biochem. Biophys.*, 115 (1966) 445.
314   J. R. Florini and C. B. Breuer, *Biochemistry*, 5 (1966) 1870.
315   G. Weber, in: E. E. Bittar and N. Bittar (Eds.), *The Biological Basis of Medicine*, Academic Press, New York, 1968, p. 263.
316   D. F. Steiner, *Vitam. Horm. (N.Y.)*, 24 (1966) 1.
317   M. E. Salas, E. Vinuela and A. Sols, *J. Biol. Chem.*, 238 (1963) 3535.
318   C. R. Morgan and J. Bonner, *Proc. Natl. Acad. Sci. (U.S.A.)*, 65 (1970) 1077.
319   M. Oravec and A. Korner, *Eur. J. Biochem.*, 27 (1972) 425.
320   D. F. Steiner and J. King, *Biochim. Biophys. Acta*, 119 (1966) 510.
321   I. G. Wool, W. S. Stirwalt, K. Kurihara, R. B. Low, P. Bailey and B. Oyer, *Recent Progr. Horm. Res.*, 24 (1968) 139.
322   J. J. Castles, F. S. Rolleston and I. G. Wool, *J. Biol. Chem.*, 246 (1971) 1799.
323   I. G. Wool and P. Cavicchi, *Biochemistry*, 6. (1967) 1231.
324   I. G. Wool and K. Kurihara, *Proc. Natl. Acad. Sci. (U.S.A.)*, 58 (1967) 2401.
325   W. S. Stirewalt, I. G. Wool and P. Cavicchi, *Proc. Natl. Acad. Sci (U.S.A.)*, 57 (1967) 1885.
326   I. G. Wool and P. Cavicchi, *Proc. Natl. Acad. Sci. (U.S.A.)*, 56 (1966) 991.
327   L. S. Jefferson, D. E. Rannels, B. L. Munger and H. E. Morgan, *Fed. Proc. Fed. Am. Soc. Exp. Biol.*, 33 (1974) 1098.
328   L. E. Gerschenson, M. B. Davidson and M. Anderson, *Eur. J. Biochem.*, 41 (1974) 139.
329   C. A. Barnett and W. D. Wicks, *J. Biol. Chem.*, 246 (1971) 7201.
330   W. D. Wicks, C. A. Barnett and J. B. McKibbin, *Fed. Proc. Fed. Am. Soc. Exp. Biol.*, 33 (1974) 1105.
331   K. L. Lee, J. R. Reel and F. T. Kenney, *J. Biol. Chem.*, 245 (1970) 5800.
332   J. R. Reel, K. L. Lee and F. T. Kenney, *J. Biol. Chem.*, 245 (1970) 5806.
333   D. E. Bushnell, J. E. Becker and V. R. Potter, *Biochem. Biophys. Res. Commun.*, 56 (1974) 815.
334   L. D. Garren, *Vitam. Horm. (N.Y.)*, 26 (1968) 119.
335   S. Castells, N. Addo and K. Kwateng, *Steroids*, 22 (1973) 171.
336   S. Castells, N. Addo and K. Kwateng, *Endocrinology*, 93 (1973) 285.
337   L. D. Garren, R. L. Ney and W. W. Davis, *Proc. Natl. Acad. Sci. (U.S.A.)*, 53 (1965) 1443.

338   M. K. Mostafapour and T. T. Tchen, *Biochem. Biophys. Res. Commun.*, 44 (1971) 774.
339   M. K. Mostafapour and T. T. Tchen, *J. Biol. Chem.*, 248 (1973) 6674.
340   S. M. el Khatib, J. Haldar and J. L. Starr, *Proc. Soc. Exp. Biol. Med.*, 143 (1973) 869.
341   S. E. Graber, M. Carrillo and S. B. Krantz, *J. Lab. Clin. Med.*, 83 (1974) 288.
342   E. Goldwasser, C. K.-H. Kung and J. Eliason, *J. Biol. Chem.*, 249 (1974) 4202.
343   S. B. Krantz, O. Gallien-Lartigue and E. Goldwasser, *J. Biol. Chem.*, 238 (1963) 4085.
344   P. P. Dukes, in: *Wirkungsmechanismen der Hormone, 18. Colloq. Ges. Physiol. Chem.*, Springer, Berlin, 1967, p. 197.
345   O. Gallien-Lartigue and E. Goldwasser, *Biochim. Biophys. Acta*, 103 (1965) 319.
346   S. B. Krantz and E. Goldwasser, *Biochim. Biophys. Acta*, 103 (1965) 325.
347   J. Paul and J. A. Hunter, *J. Mol. Biol.*, 42 (1969) 31.
348   M. Gross and E. Goldwasser, *J. Biol. Chem.*, 246 (1971) 2480.
349   M. Gross and E. Goldwasser, *Biochim. Biophys. Acta*, 287 (1972) 514.
350   G. M. Maniatis, R. A. Rifkind, A. Bank and P. A. Marks, *Proc. Natl. Acad. Sci. (U.S.A.)*, 70 (1973) 3189.
351   C. S. Chang and E. Goldwasser, *Dev. Biol.*, 34 (1973) 246.
352   C. S. Chang, D. Sikkema and E. Goldwasser, *Biochem. Biophys. Res. Commun.*, 57 (1974) 399.
353   F. Jacob and J. Monod, *J. Mol. Biol.*, 3 (1961) 318.
354   R. J. Britten and E. H. Davidson, *Science*, 165 (1969) 349.
355   R. J. Britten and D. E. Kohne, *Science*, 161 (1968) 529.
356   W. Gilbert, N. Maizels and A. Maxam, *Cold Spring Harbor Symp. Quant. Biol.*, 38 (1973) 845.
357   J. Hedgpeth, H. M. Goodman and H. W. Boyer, *Proc. Natl. Acad. Sci. (U.S.A.)*, 69 (1973) 3448.
358   C. A. Thomas Jr., R. E. Pyeritz, D. A. Wilson, B. M. Dancis, C. S. Lee, M. D. Bick, H. L. Huang and B. H. Zimm, *Cold Spring Harbor Symp. Quant. Biol.*, 38 (1973) 353.
359   G. P. Georgiev, A. Ya. Varshavsky, A. P. Ryskov and R. B. Church, *Cold Spring Harbor Symp. Quant. Biol.*, 38 (1973) 869–884.
360   *Cold Spring Harbor Symp. Quant. Biol.*, 35 (1970).
361   H. Stein and P. Hausen, *Eur. J. Biochem.*, 14 (1970) 270.
362   H. Stein and P. Hausen, in ref. 360.
363   H. Stein, H. Hameister and C. Kedinger in: E. K. F. Bautz, P. Karlson and H. Kersten, (Eds.), Regulation of Transcription and Translation in Eukaryotes, *24. Colloq. Ges. Biol. Chem.*, Springer, Berlin, 1973, p. 163.
364   K. H. Seifart, in ref. 360.
365   K. H. Seifart, P. P. Juhasz and B. Benecke, *Eur. J. Biochem.*, 33 (1973) 181.
366   D. Lentfer and A. G. Lezius, *Eur. J. Biochem.*, 30 (1972) 278.
367   K. Scherrer and J. Darnell, *Biochem. Biophys. Res. Commun.*, 7 (1962) 486.
368   K. Scherrer, H. Latham and J. Darnell, *Proc. Natl. Acad. Sci. (U.S.A.)*, 49 (1963) 240.
369   K. Scherrer and L. Marcaud, *Cell. Physiol.*, 72 (1968) 181.
370   K. Scherrer, in: E. K. F. Bautz, P. Karlson and H. Kersten (Eds.), Regulation of Transcription and Translation in Eukaryotes, *24. Colloq. Ges. Biol. Chem.*, Springer, Berlin, 1973 p. 81.
371   J. Niessing and C. E. Sekeris, *Biochim. Biophys. Acta*, 209 (1970) 484.
372   R. W. Shearer and B. J. McCarthy, *Biochemistry*, 6 (1967) 283.
373   L. Lim and E. D. Canellakis, *Nature (London)*, 227 (1970) 710.
374   J. Kates, in ref. 360.
375   M. P. Edmonds, M. H. Vaughan and H. Nakazato, *Proc. Natl. Acad. Sci. (U.S.A.)*, 68 (1971) 1336.
376   Y. Lee, J. Mendecki and G. Brawerman, *Proc. Natl. Acad. Sci. (U.S.A.)*, 68 (1971) 1331.

377   J. E. Darnell, L. Philipson, R. Wall and M. Adesnik, *Science*, 174 (1971) 507.
378   J. Niessing and C. E. Sekeris, *Nature (London)*, *New Biol.*, 243 (1973) 9.
378a R. P. Perry and D. E. Kelley, *Cell*, 1 (1974) 37.
379   R. Haselkorn and L. B. Rothman-Denes, *Annu. Rev. Biochem.*, 42 (1973) 397.
380   H. Kroeger, *Nature (London)*, 200 (1963) 1234.
381   H. Kroeger, *Exptl. Cell Res.*, 41 (1966) 64.
382   H. Kroeger and M. Lezzi, *Annu. Rev. Entomol.*, 11 (1966) 1.
383   M. Lezzi and M. Robert, in: W. Beermann (Ed.), *Results and Problems in Cell Differentiation*, Vol. 4, Springer, Berlin, 1972, p. 35.
384   U. Clever, *Chromosoma*, 17 (1965) 309.
385   H. D. Berendes, in: W. Beermann (Ed.), *Developmental Studies on Giant Chromosomes*; *Results and Problems in Cell Differentiation*, Vol. 4 Springer, Berlin, 1972, p. 181.
386   U. Clever, *Annu. Rev. Genet.*, 2 (1968) 11.
387   M. Ashburner, *Adv. Insect Physiol.*, 7 (1970) 2.
388   M. Lezzi and L. I. Gilbert, *Gen. Comp. Endocrinol. Suppl.* 3 (1972) 159.
389   W. Beermann (Ed.), *Results and Problems in Cell Differentiation*, Vol. 4, Springer, Berlin, 1972.
390   K. Marushige and J. Bonner, *J. Mol. Biol.*, 15 (1966) 160.
391   H. D. Berendes, F. M. A. van Breugel and T. K. H. Holt, *Chromosoma*, 16 (1965) 35.
392   W. C. Claycomb, R. E. LaFond, and C. A. Villee, *Nature (London)*, 234 (1971) 302.
393   L. F. Congote, C. E. Sekeris and P. Karlson, *Exp. Cell Res.*, 56 (1969) 338.
394   L. D. Garren, R. R. Howell, G. M. Tomkins and R. M. Crocco, *Proc. Natl. Acad. Sci. (U.S.A.)*, 52 (1964) 1121.
395   E. B. Thompson, D. K. Granner and G. M. Tomkins, *J. Mol. Biol.*, 54 (1970) 159.
396   G. M. Tomkins, T. D. Gelehrter, D. Granner, D. W. Martin Jr., H. H. Samuels and E. B. Thompson, *Science*, 166 (1969) 1474.
397   G. M. Tomkins, B. B. Levinson, J. D. Baxter and R. Dethlefsen, *Nature (London)*, *New Biol.*, 239 (1972) 9.
398   M. Singer and P. Leder, *Annu. Rev. Biochem.*, 35 (1966) 195.
399   G. A. Stewart and E. Farber, *J. Biol. Chem.*, 243 (1968) 4479.
400   F. T. Kenney, K. L. Lee, C. D. Stiles and J. E. Fritz, *Nature (London)*, *New Biol.*, 246 (1973) 208.
401   S. Penman, M. Rosbash and M. Penman, *Proc. Natl. Acad. Sci. (U.S.A.)*, 67 (1970) 1878.
402   R. D. Palmiter and R. T. Schimke, *J. Biol. Chem.*, 248 (1973) 1502.
403   G. P. Georgiev, *J. Theoret. Biol.*, 25 (1969) 473.
404   C. Coutelle, A. P. Ryskov and G. P. Georgiev, *FEBS Letters*, 12 (1970) 21.
405   G. P. Georgiev, A. P. Ryskov, C. Coutelle, V. L. Mantieva and A. R. Avakyan, *Biochim. Biophys. Acta*, 259 (1972) 259.
406   J. Paul, *Nature (London)*, 238 (1972) 444.
407   F. H. C. Crick, *Nature (London)*, 234 (1971) 25.
408   V. G. Allfrey, C. S. Teng and C. T. Teng, in: D. W. Ribbons, J. F. Woessner and J. Schultz (Eds.), *Nucleic Acid–Protein Interaction—Nucleic Acid Synthesis in Viral Infection*, North Holland, Amsterdam, 1971, p. 144.
409   C. H. Waddington, *Science*, 166 (1969) 639.

Chapter II

Hormone-Sensitive Adenyl Cyclase Systems: Properties and Function

THEODOR BRAUN AND LUTZ BIRNBAUMER

Department of Physiology, Northwestern University Medical Center, Chicago, Ill. (U.S.A.)

1. Introduction:
present patterns of hormone action in mammalian cells

The fundamental contribution of Sutherland and his associates to our understanding of hormone action is the recognition of the general role of cyclic AMP (cAMP) as a mediator of hormonal response[1–5]. Evidence accumulated during the last decade has established that catecholamines and a number of peptide, protein and glycoprotein hormones act *via* the adenyl cyclase–cAMP system in various tissues[5]. Similarly with the pioneering and persistent work of scientists, such as Jensen[6,7], Gorski[8], Edelman[9], Liao[10], Baulieu[11], Williams-Ashman[12], Tomkins[13,14], O'Malley[15], Mueller[16], and many others, the mode of action of steroid hormones at the cellular and subcellular level has been elucidated. Most hormones can be classified in two groups, based on their mode of action. One group, to which all steroid hormones belong, includes those which are able to pass through the cell's plasma membrane; and ultimately affect gene expression. The second group of hormones, to which catecholamines, peptide, protein and glycoprotein hormones belong, does not enter target cells, but involves an interaction with receptors of the cell membrane. Fig. 1 illustrates our present understanding of the sequence of events, from the initial reaction of the hormone with its target cell to the final expression of hormonal activity. Steroid hormones enter the target cell and bind to cytosol receptor proteins. Subsequently, the steroid–receptor complex is "modified" by a temperature-dependent process and translocated to the nucleus. In the nucleus the

References p. 101

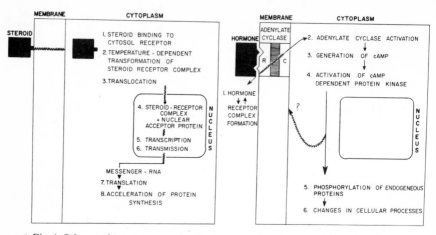

Fig. 1. Scheme of current concepts of hormone action in mammalian target cells.

steroid–receptor complex binds to specific acceptor sites on the chromatin and by a mechanism yet unknown promotes the synthesis of messenger-RNA. The messenger-RNA transmitted to the cytoplasm carries the information required for *de novo* protein synthesis.

In the case of catecholamines and certain peptide, protein and glyco-protein hormones the regulatory message is relayed intracellularly by cAMP, which is generated by the membrane-associated adenyl cyclase system. In this case the hormone interacts with receptors of the plasma membrane. Then the catalytic component of the adenyl cyclase system is activated to generate cAMP through a process designated as coupling[17]. Cyclic AMP, so formed, initiates a chain of events leading to phosphorylation by cAMP-dependent protein kinases which ultimately results in the expression of hormonal activity. It has been suggested that all effects attributed to cAMP in mammalian cells are evoked by the activation of cAMP-dependent protein kinases[18,19]. Thus, information transfer between the hormone, whether steroid or non-steroid, and its target cell follows a sequence involving complex effector systems. In this review we shall only examine the action of hormones involving cAMP as an intracellular messenger.

In this review, in discussing the nature of the adenyl cyclase system, we shall address ourselves to the question: How does the initial reaction of the hormone with a receptor on the plasma membrane lead to adenyl cyclase activation?

2. Characterization and properties of adenyl cyclase systems

(a) General

Adenyl cyclase systems catalyze the formation of cAMP from ATP.

$$ATP \xrightarrow[Mg^{2+}]{Adenyl\ cyclase} cAMP + PP_i$$

The presence of adenyl cyclase systems has been demonstrated in bacteria[20-23], slime molds[24,25], hepatic flukes[26,27], worms[28], insects[28], fish[29], amphibians[30,31], reptiles[32,33], birds[34], as well as in mammals[35,36]. On the basis of circumstantial evidence from studies using detergents and proteolytic and lipolytic enzymes, it is widely believed that adenyl cyclase systems consist of lipid and protein entities which form a complex, intimately associated with the plasma membrane. Adenyl cyclases from cells of higher organisms including mammals have resisted for a long time attempts at purification. Recently adenyl cyclase from a number of rat tissues has been solubilized with Lubrol-PX, a nonionic detergent[37-39]. Hormonal sensitivity of the solubilized adenyl cyclase is restored after removing the detergent[39], or by the addition of specific phospholipids[37]. Adenyl cyclases from *Brevibacterium liquefaciens*[40,41] and *Escherichia coli*[42,43] have been solubilized and extensively purified. However, bacterial adenyl cyclases are hormone-insensitive and hence we shall not deal with them in this review. The reader is referred to excellent reviews on the role of cAMP in bacteria by Pastan and Perlman[23], and Pastan *et al.*[44].

(b) Cellular distribution and localization of adenyl cyclase systems in the plasma membrane

Adenyl cyclase activity of the cell is in most part associated with the plasma membrane. However, adenyl cyclase activity also has been found in the mitochondrial fractions of adrenal[45] and cerebral cortex[46] and associated with the nuclei of liver[47], and prostate[48], and with the sarcoplasmic reticulum of heart[49] and skeletal[50,51] muscle. These latter findings suggest that adenyl cyclases may be components of intracellular organelles. However, since these data might reflect homogenization artifacts, a definitive conclusion in this respect must wait until appropriate methods are developed to localize unequivocally the presence of adenyl cyclase systems in intracellular compartments of intact cells. There is evidence that the receptors of the adenyl cyclase system are located at or oriented toward the outer surface of the plasma membrane, while the catalytic component with its active center is oriented toward or located at the inner surface of the plasma membrane. In the studies by Øye and Sutherland[52] treatment of intact turkey erythrocytes with proteolytic enzymes decreased the activity of membrane-bound ATPase without a corresponding change in the activity of adenyl cyclase. Based on the fact that the ATP on the outside was hydrolyzed by intact erythrocytes, they assumed that the active site of ATPase is located at or near the external surface of the plasma membrane. However, when hemolyzed erythrocytes were treated with proteolytic enzymes, both ATPase and adenyl cyclase activities were destroyed. In the studies by Rodbell et al.[53] treatment of isolated fat cells with trypsin led to alterations of receptors and changes in hormonal stimulation of the adenyl cyclase system, but did not affect the activity of the catalytic unit. In intact cells apparently only the receptors at or near the outer surface of the plasma membrane are accessible to trypsin. However, when isolated membrane preparations were exposed to digestion by trypsin the activity of the catalytic component was abolished[53].

(c) Substrate specificity and metal requirement for catalytic activity of adenyl cyclase systems

Adenyl cyclases use ATP as substrate. It has been demonstrated that dATP can substitute for ATP in rat fat cell ghost[54] and tadpole and frog erythrocyte ghost preparations[30]. The ATP analog AMP–PNP in which the oxygen of the β–γ pyrophosphate bond has been replaced by nitrogen, is also used as substrate by adenyl cyclases in rat liver[55], human platelets[56] and kidney membranes. Thus it appears that the catalytic unit of the adenyl cyclase system exhibits nucleotide base specificity. Mg^{2+} was found originally to be essential for the catalytic activity of the liver adenyl cyclase[57]. Most likely, the true substrate for the catalytic unit of the cyclase system is $Mg\cdot ATP^{58,59}$. Stochiometrically at least a $1:1$ ratio of $Mg\cdot ATP$ is required, since free ATP inhibits adenyl cyclase activity[58]. Mn^{2+} (refs. 28, 58–61) and partially Co^{2+} (refs. 58, 59) substitute for Mg^{2+} in adenyl cyclase systems presently studied, but not Ca^{2+}, Zn^{2+}, Hg^{2+} or Cu^{2+}, which generally inhibit adenyl cyclase activity[58,62]. The exception to this pattern is the bull sperm adenyl cyclase system in which Mg^{2+} can be effectively replaced by Mn^{2+}, Co^{2+}, Zn^{2+} and Cd^{2+}, and partially by Ca^{2+} (ref. 61).

(d) Effect of fluoride ion

Adenyl cyclase systems in preparations of broken cells from multicellular organisms are stimulated by fluoride ion[63]. Again, the exception to this pattern is the adenyl cyclase system in broken-cell preparations of rat epididymal spermatozoa[61]. The exact nature of the fluoride effect upon adenyl cyclase systems is not yet known. Observations made in fat cell ghosts[58] and liver plasma membranes[64] indicate that fluoride and hormones stimulate the same catalytic unit of adenyl cyclase systems, through different, but closely associated mechanisms[64,65]. Fluoride ion has been shown under appropriate conditions to interfere with hormonal stimulation[65].

(e) Effect of hormones

Adenyl cyclase systems in higher organisms are hormonally regulated. The various hormones of disparate structures shown to influence adenyl cyclase activity are listed in Table I. In some tissues the cyclase system is stimulated by a single hormone, e.g. adrenal cyclase by ACTH, and thyroid cyclase by TSH. In other tissues with multiple-hormone responsiveness, each hormone-specific adenyl cyclase system is anatomically separated, localized in different cell types. For example, in the kidney the adenyl cyclase in cortical cells is stimulated by PTH, while that in the medullary cells is stimulated by VP and OT. Similarly, in testes the adenyl cyclase system in interstitial Leydig cells is activated by LH but that in seminiferous tubule cells by FSH[130,133]. In some other tissues, such as myocardium, liver and adipose tissue two or more structurally unrelated hormones stimulate the same adenyl cyclase (Table I). It has also been established that in fat cells the ability of hormones to stimulate adenyl cyclase varies profoundly from one species to another[110,113].

Catecholamines influence the activity of adenyl cyclase systems in a variety of tissues. The diverse effects of catecholamines in various tissues and cell types[143] have been classified as α- or β-effects. On this basis, it is assumed that two types of adrenergic receptors exist in the various cell types, or perhaps, within the same cell. The effect of catecholamines can be either stimulating or inhibiting. The stimulatory effect of catecholamines upon adenyl cyclase systems has been linked to the adrenergic receptors[86]. Furthermore, the stimulation has been demonstrated in broken-cell preparations, of both tissues and isolated cells, indicating a direct influence upon

TABLE I

Adenyl cyclase systems in tissues and cells of higher organisms influenced by hormones

Tissue or cells	Hormone	Effect on adenyl cyclase system	Reference
Erythrocytes	Catecholamines (β)	Stimulation	52, 66–68
Blood platelets	Catecholamines (α)	Inhibition	69
Melanocytes	MSH	Stimulation	
	Catecholamines (α)	Inhibition	70, 71
Parotid gland	Catecholamines (β)	Stimulation	72, 73
Pineal gland	Catecholamines (β)	Stimulation	74–76
Spleen	Epinephrine	Stimulation	77
Lung	Epinephrine	Stimulation	77
Liver	Catecholamines (β), glucagon	Stimulation	78–85
Heart musle	Catecholamines (β), glucagon, T_4	Stimulation	78, 86–89
	acetylcholine	Inhibition	
Skeletal muscle	Catecholamines (β)	Stimulation	90–93
Smooth muscle	Catecholamines (β)	Stimulation	94, 95
	Catecholamines (α)	Inhibition	
Brain	Catecholamines (β), serotonine	Stimulation	77, 96–100
Pituitary	Hypothalamic releasing hormones	Stimulation	101, 102
Pancreas	Glucagon, catecholamines (β)	Stimulation	103, 104
	catecholamine (α)	Inhibition	
Adipose tissue:	Catecholamines (β) ACTH, glucagon,		
(white) (rat)	LH, TSH, secretin	Stimulation	53, 58, 105–1
(rabbit)	ACTH, α-, β-MSH, β-LPH, glucagon		
	catecholamines (β)	Stimulation	110,111
(human)	Catecholamines (β)	Stimulation	112, 114
	catecholamines (α)	Inhibition	
Adipose tissue:	Catecholamines	Stimulation	115, 116
(brown) (rat)			
Kidney	PTH, calcitonin, OT, VP	Stimulation	117–120, 56
Bone	PTH, calcitonin	Stimulation	121–123
Ovary	LH	Stimulation	124, 125
Testis	FSH, LH, hCG	Stimulation	126–133
Adrenal	ACTH	Stimulation	134–139, 45
Thyroid	TSH	Stimulation	77, 134–142

Abbreviations: MSH = melanocyte-stimulating hormone; T_4 = thyroxine; ACTH = adrenonocorticotropin; L
luteinizing hormone; TSH = thyroid-stimulating hormone; LPH = lipotropic hormone; PTH = parathorm
OT = oxytocin; VP = vasopressin; FSH = follicle-stimulating hormone; hCG = human choriogonadotropin.

adenyl cyclase activity. Whether or not the inhibitory effect on adenyl cyclase activity is similarly mediated through direct influence on the cyclase systems remains unclear. It has been observed in several intact tissues and cells that α-adrenergic responses are accompanied by a fall in the intracellular level of cAMP[144]. Inhibition of adenyl cyclase activity in broken-cell preparations has been observed only recently. Triner et al.[145] using whole homogenate of rat uterus have found that epinephrine in the presence of a β-adrenergic blocking agent inhibits adenyl cyclase activity, whereas in the absence of the blocking agent adenyl cyclase stimulation is observed. Recently in rabbit fat-cell membrane preparations, inhibitory response to epinephrine has been unmasked using β-adrenergic blocking agents (Table II). These findings suggest the possibility that α-adrenergic effects, might be mediated through a direct inhibitory influence, at least in certain cell types, on adenyl cyclase systems.

TABLE II

Effect of epinephrine in the absence and presence of DCI on rabbit fat-cell ghost adenyl cyclase activity

Epinephrine concentration $(M \cdot 10^{-6})$	Adenyl cyclase activity (cAMP pmoles/mg protein/min formed)	
	Without	With DCI[a]
—	76.1	76.8
1.5	81.9	58.6
5	90.5	63.8
50	108.0	79.8
150	93.5	92.7
500	107.0	115.0

[a] DCI (dichlorisoproterenol-HCl) at a constant dose of $4 \cdot 10^{-5}$ M was used.
Fat-cell ghosts were prepared from pooled perirenal adipocytes.
Adenyl cyclase activity was assayed using conditions described by Bär and Hechter[165].

(f) Effect of insulin

Insulin has been shown to lower cAMP levels in alloxan-diabetic rat liver, or normal livers in the precence of glucagon or catecholamines[146,147]. In rat adipose tissue, insulin lowered basal levels as well as glucagon and ACTH-augmented cAMP levels[148]. The idea that insulin lowers cAMP levels by inhibiting adenyl cyclase activity has been tested in fat-cell ghosts[149] and liver-membrane preparations[150]. Until recently, attempts to demonstrate direct inhibitory effects on adenyl cyclase systems in isolated membrane preparations have failed. However, Illiano and Cuatracasas[151], and Hepp[152,153] have reported inhibition of basal and glucagon-stimulated adenyl cyclase activity in isolated rat-liver membranes. In these studies inhibition by insulin was obtained at low concentrations (5–100 μU/ml) but not with higher concentrations of the hormone. Inhibitory effects of insulin on adenyl cyclase have also been observed by Jimenez de Asua et al. in cultured fibroblasts[154]. In this study the dose–response curve for insulin extended over 4 orders of magnitude indicating that the mechanism by which this effect is exerted must be complex, involving cooperative interactions. In our own studies and those of others[149,155,156], insulin was found ineffective in inhibiting the rat fat-cell ghost adenyl cyclase over a wide range of hormone concentrations. It is still unclear why the previous attempts to influence adenyl cyclase activity in cell-free systems were unsuccessful.

3. The nature of hormone selectivity of adenyl cyclase systems

Adenyl cyclase systems in the various cell types are hormone-selective. Within recent years it became clear that the hormone selectivity of adenyl cyclase systems resides in their receptors which "recognize" and combine with the hormone. Experimental evidence for this conclusion comes mainly from studies exploring the nature of hormone selectivity of fat-cell adenyl cyclase[105,107]. The rat fat-cell adenyl cyclase activity is modified by a number of structurally unrelated hormones, such as ACTH, epinephrine, glucagon, secretin, TSH and LH[53,58,105,107]. The question then is, whether each of the hormones interacts with a separate adenyl cyclase or if all the hormones activate a single unit of adenyl cyclase through distinct receptors. In the fat-cell membrane system the stimulatory effects by supramaximal doses of one hormone with the simultaneous addition of other hormones in various combinations were not additive, indicating that a single catalytic unit of adenyl cyclase is stimulated by the various hormones[105,108]. Yet, it has also been shown that the stimulatory effect of ACTH is selectively inhibited by a structurally related ACTH analogue[107], while that of epinephrine is only blocked by β-adrenergic blocking agents[105]. These findings suggested that receptor sites for these hormones are distinct. This conclusion was further supported by the finding that hormone-receptors exhibit differential accessibility and susceptibility to proteolytic enzymes. Thus, when fat cells were pre-incubated with trypsin, adenyl cyclase in ghosts did not respond to glucagon, the stimulatory effects of ACTH and secretin were reduced by 40–60% respectively, but the stimulatory effects of epinephrine and fluoride were unaffected[53]. It has also been shown that ACTH stimulation of rat fat-cell adenyl cyclase has a specific calcium requirement, while that of other hormones does not[105,107].

Moreover, ACTH stimulation of the rat fat-cell plasma membrane adenyl cyclase is uniquely influenced by glucocorticoid levels, i.e., it is reduced following adrenalectomy or hypophysectomy and is enhanced by dexamethasone treatment (see below).

In conclusion, these findings provide experimental background for the view that hormone selectivity in the adenyl cyclase systems depends upon distinctive receptors which are coupled to a single catalytic unit in the membrane.

4. Factors modifying hormonal sensitivity of adenyl cyclase systems

Several factors characteristically influence the cyclase system's responsiveness to a specific hormone, independent of changes in basal activity. Studies of these factors provide strong evidence that hormone receptors and the catalytic unit of adenyl cyclase systems are under separate control.

(a) Developmental factors

Rosen and Rosen[30,66], and Rosen and Ehrlichman[67] were the first to show that although adenyl cyclase catalytic activity was already present in hemolyzates of frog and tadpole erythrocytes, epinephrine-responsiveness was acquired at a certain stage of metamorphosis of the amphibian. Adenyl cyclase stimulation in erythrocyte hemolyzates was first demonstrated during natural or thyroxin-induced metamorphosis in tadpoles which had developed front legs and were undergoing the process of tail resorption[67]. Studies of adenyl cyclases in broken cell preparations of rat liver[82], pineal gland[75], and testis[129] and in rat-brain slices[97] have produced similar findings, demonstrating the development of hormonal-sensitivity independent of the development of the catalytic unit of the respective cyclase systems.

Studies by Robison et al.[97], revealed that adenyl cyclase response to fluoride in broken-cell preparations of rat brain and norepinephrine-response in brain slices was acquired at different times postnatally. Thus, in addition to the fact that fluoride and hormonal stimulation of the cyclase catalytic component is accomplished through distinct processes[64], it appears that separate developmental factors determine their emergence.

(b) Steroid hormones

Recent studies have drawn attention to the fact that steroids can selectively influence hormonal-sensitivity of cyclase systems in several tissues including adipose tissue[109], liver[82] and pineal gland[75]. The ability of ACTH to specifically stimulate adenyl cyclase in fat-cell ghosts is markedly reduced in adrenalectomized or hypophysectomized rats compared to that in sham-operated control rats (Table III). In contrast to ACTH, the stimulatory effects of epinephrine, glucagon and fluoride upon adenyl cyclase activity of ghost preparations from adrenalectomized[109,157] or hypophysectomized[109] rats were unchanged.

Responsiveness to ACTH in fat-cell ghosts from adrenalectomized rats

TABLE III

Effect of ACTH on fat-cell ghost adenyl cyclase activity from adrenalectomized or hypophysectomized rats relative to that from sham-operated rats

Experiment	Days after Surgery	Relative ACTH effect (% of maximal effect in corresponding control groups)
1	6	21
2	7	16
3	7	12
4	12	19
5	23	16
6	24	35
		Adrenalectomized $\bar{x} = 19.8$
1	6	36
2	18	16
3	20	20
4	100	12
5	116	17
		Hypophysectomized $\bar{x} = 20.2$

Fat-cell ghosts were prepared from pooled epididymal adipocytes.

was restored, or even increased to supernormal levels, by treatment of adrenalectomized rats with the synthetic glucocorticoid, dexamethasone. The degree of restoration of ACTH sensitivity induced by dexamethasone depended on the time course of steroid treatment. Thus dexamethasone administered over 5–8 h increased ACTH response significantly but not to normal levels; the same total dose of dexamethasone administered over 52 h produced supernormal sensitivity[109]. Prolonged treatment of 40–60 h with dexamethasone increased ACTH sensitivity to supernormal levels not only in adrenalectomized rats but in sham-operated rats as well. In both groups the dexamethasone effect is specific for ACTH; the responses to epinephrine, glucagon and fluoride are unchanged (Fig. 2).

The same restorative effects of dexamethasone treatment on ACTH response are obtained in fat-cell ghost adenyl cyclase system from hypophysectomized rats[109]. Cortisol has similar, though less pronounced effects on the ACTH-stimulated adenyl cyclase in the ghost preparations; 11-deoxycorticosterone was ineffective. The effect of dexamethasone in augmenting ACTH stimulation was blocked by simultaneous administration of either actinomycin D, which is known to inhibit DNA-dependent RNA synthesis, or cycloheximide, which is known to inhibit protein synthesis. Neither actinomycin D nor cycloheximide influenced the response of fat-cell membrane adenyl cyclase to epinephrine or fluoride. The finding that actinomycin D and cycloheximide block the effect of dexamethasone to selectively increase the response to ACTH strongly suggests that glucocorticoids regulate the synthesis of a molecular entity required for ACTH action. The induction of ACTH sensitivity in rat fat membranes appears to be similar, in principle, to the induction of adaptive enzymes in mammalian cells by adrenal steroid hormones. The induction of specific hepatic enzymes by glucocorticoids has been shown to be an increase in the rate of enzyme synthesis and to be prevented by actinomycin D. On the assumption that glucocorticoids act at the gene locus, it would appear that glucocorticoids modify the degree of expression of the gene involved in ACTH sensitivity of the rat fat-cell membrane. Thus, gene expression appears to be markedly reduced after adrenalectomy or hypophysectomy due to glucocorticoid deficiency. Since dexamethasone induces supernormal sensitivity to ACTH in intact as well as adrenalectomized rats, the gene seems to be partially expressed in normal rats, and can be more fully expressed when glucocorticoid levels are increased above normal.

The molecular nature of the entity required for ACTH stimulation of

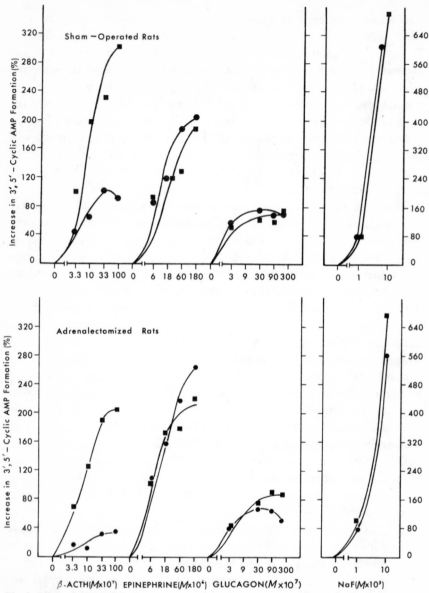

Fig. 2. Effect of dexamethasone treatment of sham-operated and adrenalectomized rats on the response of ghost cyclase to ACTH, epinephrine, glucagon, and NaF. —●—, control, placebo-treated groups; —■—, dexamethasone-treated groups. Dexamethasone (total dose 0.5 mg/100 g body weight) was injected intraperitoneally in five separate injections over 52 h. Data from Braun and Hechter[109].

rat fat-cell ghost adenyl cyclase induced by glucocorticoids has not been established. It could be ACTH receptors or a factor required for coupling the primary reaction of ACTH with receptor to adenyl cyclase.

Modification of catecholamine sensitivity of adenyl cyclase systems by steroids has been demonstrated in broken-cell preparations of liver and pineal gland. Following adrenalectomy or hypophysectomy in liver homogenates and particulate fractions, Bitenski et al.[82], have observed marked and selective augmentation of adenyl cyclase response to epinephrine but not to glucagon. Administration of prednisolone during, or several days following surgery, prevented the rise in the stimulatory effect of epinephrine upon liver adenyl cyclase. After prolonged administration of ACTH to hypophysectomized or intact weanling rats, prednisolone produced a similar decrease in the activity of epinephrine-responsive liver adenyl cyclase. Additionally, Bitenski et al.[82] have noted that epinephrine-responsive adenyl cyclase activity in liver broken-cell preparations is significantly higher in the female than in the male rats. Testosterone administration to rats of both sexes resulted in a marked decline in the stimulatory effect of epinephrine upon liver adenyl cyclase.

Crayton and Weiss have observed that stimulation of adenyl cyclase activity in broken-cell preparations of the pineal gland by norepinephrine was significantly greater in male than in female rats, although basal activities were similar[76]. Adenyl cyclase activity in pineal gland homogenates from female rats in estrus, metestrus and diestrus were clearly stimulated by norepinephrine, but not in preparations from rats in proestrus[75]. Estrogens have been implicated in the unresponsiveness of pineal adenyl cyclase activity to norepinephrine during proestrus. If ovariectomized rats are treated for 59 h with estradiol, the stimulatory effect of norepinephrine upon pineal gland adenyl cyclase is reduced to 80%. Estradiol administered to ovariectomized rats 1 h before the assay had no inhibitory effect on the norepinephrine response. Estradiol added directly to the assay mixture also failed to inhibit the stimulatory effect of norepinephrine on pineal adenyl cyclase[75].

Whether the effect of steroids upon epinephrine-responsive liver or norepinephrine-responsive pineal gland adenyl cyclases also involves protein synthesis has not been determined.

The above findings together with the ontogenetic studies mentioned before, serve to illustrate the complex molecular nature of the adenyl cyclase systems, where the molecular entities associated with hormone-selectivity are induced and maintained by distinct genetic and hormonal factors. These studies implicate different steroids as factors controlling hormonal-sensitivity, possibly through their effect on synthesis of a protein component of the adenyl cyclase system. It should be pointed out that the *steroid* and not the cyclase-stimulating hormone appears to be the inducer for the hormone-selectivity unit, in contrast to the well known phenomena of induction of antibody synthesis by a specific antigen or substrate-induced enzyme synthesis.

(c) Species and tissue variation factors

There are differences in the sensitivity of adenyl cyclase systems in similar tissues of various species to the same hormone(s)[110]. For example, ACTH was found to be effective in stimulating adenyl cyclases of the rat and rabbit fat-cell ghost preparations. However, the concentrations of α-ACTH$^{1-39}$ and β-ACTH$^{1-24}$ required for half-maximal stimulation of the rat fat-cell ghost adenyl cyclase are ten times higher than that required for activation of the rabbit cyclase system (Table IV). There are also differences in the sensitivity of adenyl cyclase systems to the same hormone among different tissues in the same animals. From Table V, it appears that the concentrations of glucagon required for half-maximal binding and adenyl cyclase activation in fat-cell ghosts are about 23–30 times higher than those in liver or pancreatic tissue preparations. It appears that the glucagon-sensitive adenyl cyclase of the rat fat-cell ghosts possesses only high-affinity receptors, while the liver and pancreatic tissue cyclase systems possess high and low affinity receptors as well. Hence, the possibility arises

TABLE IV

Comparison of the effect of α_sACTH$^{1-39}$ and ACTH$^{1-24}$ on adenyl cyclase activity in rat and rabbit fat-cell ghost

Hormone	Rat		Rabbit	
	K_m (M)	Maximal effect[a]	K_m (M)	Maximal effect[a]
ACTH$^{1-39}$	$2.4 \cdot 10^{-6}$	2.5	$2.7 \cdot 10^{-7}$	4.5
ACTH$^{1-24}$	$1.3 \cdot 10^{-6}$	3.2	$1.3 \cdot 10^{-7}$	5.8

[a] Ratio of hormone effect to basal. Data from Hechter and Braun[17] (1971).

that variability in hormonal response, at least among different tissues may be due to differences in the binding affinities of receptors of the respective adenyl cyclase systems. This does not preclude differences in the effectiveness of the coupling process. This possibility has yet to be explored.

TABLE V

Comparison of glucagon-binding and adenyl cyclase activation in broken-cell preparations from liver, pancreas and adipocytes

Tissue	Binding (K_m)	Adenyl cyclase activation (K_m)	Maximal effect[a]	Reference
Liver	$4 \cdot 10^{-9}$	$4 \cdot 10^{-9}$	10–16	181
Pancreas	$1.4 \cdot 10^{-9}$	$2.8 \cdot 10^{-9}$	2.5	104
Fat-cell ghosts	$1.5 \cdot 10^{-7}$	$1 \cdot 10^{-7}$	2	107
		$7 \cdot 10^{-8}$	3	105

For the preparation of this table data from the cited studies were evaluated. In these studies liver plasma membranes, a particulate fraction obtained from homogenate of pancreatic tissue by centrifugation at $10000 \times g \times 10$ min, and fat-cell ghosts from rats were used.
[a] Ratio of hormone effect to basal.

5. Methodological approaches to study hormone-responsive adenyl cyclase systems

Hormonal effects upon adenyl cyclase systems can be studied in the whole animal, in isolated, relatively intact tissues, and in cells or broken-cell preparations. The various preparations offer advantages and limitations. Studies using intact cell preparations, though relatively complex, are more directly relevant to physiological conditions. When precise analysis of hormone actions at the molecular level is the goal, broken-cell preparations are unavoidable. Studies with intact animals and intact-cell preparations have established the physiological significance of the adenyl cyclase–cAMP system as a mediator of hormonal response. The hormonal effect on adenyl cyclases in intact cells is determined by evaluating changes in intracellular levels of cAMP[91,158]. Changes in cAMP levels thus may result from direct influence of the hormone on adenyl cyclase, or from changes in the rate of cAMP hydrolysis by intracellular phosphodiesterases. To examine the activity of an adenyl cyclase system, methylxanthines are usually used to inhibit phosphodiesterase activity[159,160]. The observation that the hormone and the phosphodiesterase inhibitor act synergistically in intact cells is generally considered evidence that the hormone acts by stimulating adenyl cyclase. Changes in cAMP levels in tissues and cells can be determined by the recently developed prelabeling technique[161,162]. This procedure is based on prelabeling the intracellular ATP pools with radioactive precursors (adenine, adenosine) which readily enter the cell. The labeled ATP pools serve as a precursor for the cAMP.

In recent years hormone-responsive adenyl cyclase systems have been extensively studied in isolated plasma membrane preparations. The advantage of studying adenyl cyclases in isolated membranes is that it allows measurement of the initial step of hormone-binding, as well as the subsequent activation of the adenyl cyclase. Furthermore, in isolated membrane preparations it is possible to determine the site of action of various agents (*e.g.*, nucleotides, phospholipids, chelating agents, calcium ions, etc.), whether affecting binding or the coupling process. Hormone binding is quantitated by measuring the amount of membrane-bound radioactively labeled hormone. It is necessary to determine whether the hormone binding is specific, *i.e.*, whether it is displaced by the native, but not by other structurally unrelated hormones. Attention must be paid to the composition of the incubation medium in which hormone-binding is performed, since chelating agents, nucleotides, agents used to regenerate ATP, might influence the amount of binding and the binding constants[163]. And, since it is the objctive that hormone-binding is to be performed at steady-state hormone levels, one must ascertain that the labeled hormone is not degraded in the cell-free system.

Adenyl cyclase activity in plasma-membrane preparations is determined by measuring accumulation of ^{32}P-cAMP from α^{32}P-ATP[164,165]. Isolated membrane preparations even if purified contain a high level of ATP-hydrolyzing enzymes and cAMP phosphodiesterase. To avoid ATP hydrolysis by ATPases an ATP-regenerating system is included in the assay mixture[165]. Recently an ATP analog, AMP–PNP, which is not hydrolyzed by ATPases, has been introduced as substrate in assays measuring adenyl cyclase activity[55]. Degradation of the ^{32}P-cAMP formed is protected by methylxanthines[164] or by an excess of unlabeled cAMP[165]. However, as in the case of binding studies, the presence of these agents in the assay mixture might influence hormonal sensitivity, as well as the kinetics of adenyl cyclase activation[163].

The coupling process is operationally defined as the process linking the primary step of hormone–receptor complex formation to adenyl cyclase activation. This process is evaluated indirectly by analyzing the effect of various substances which modify adenyl cyclase activity without influencing the initial binding step.

6. Molecular aspects of hormone–receptor interactions in adenyl cyclase systems

A hormone receptor has two functions: (*a*) discrimination for a specific hormone and (*b*) generation of a response. In the following sections we shall analyze the characteristics of hormone receptors both in intact-cell systems to gain insight into their *in situ* requirements, and in isolated membrane systems. We shall (*a*) evaluate whether or not isolated membrane systems are adequate models for studying both the discriminatory and the signal-generating functions of hormone receptors, (*b*) analyze the utility, restrictions and meaning of hormone-binding studies, and (*c*) discuss what we know about the so-called coupling process. Since extensive studies have been made on the mechanism of action of glucagon, both in the intact liver and in isolated liver plasma membranes, we shall rely heavily on evidence stemming from these studies. Whenever appropriate, however, we shall also discuss evidence from other model systems dealing with other tissues and with other adenyl cyclase-activating hormones.

(a) Characteristics of adenyl cyclase activation in intact cells

The ultimate test of physiological significance of properties of cell-free systems is to show that these properties are similar to those observed with intact cells. Accordingly, the hormonal response of adenyl cyclases in cell-free systems should exhibit the same specificity and similar dose–response relationships and kinetics as seen in intact cells. The characteristics of the glucagon-stimulated adenyl cyclase system in intact liver have been explored by Miller, Exton and others[166–170], by measuring either cAMP levels or the cellular effect of cAMP on glycogenolysis and gluconeogenesis. They have shown that: (*a*) Among several peptides tested, only glucagon affects the system. Secretin, structurally very similar to glucagon[172], does not stimulate cAMP production; and des-1-histidine glucagon (DH-glucagon), is inactive in producing hyperglycemia in the dog[173,174]; (*b*) the range of concentration over which glucagon acts to increase intracellular levels of cAMP in the perfused liver is $5 \cdot 10^{-11}$ to about 10^{-7} M;

(c) allowing for the diffusion of the hormone to liver cells, the effects of glucagon on cAMP production seem to be established very rapidly with no significant lag before achieving maximal rates; (d) administration of glucagon followed by withdrawal results in a rapid fall in cAMP levels in the perfused liver; half-maximal levels are observed within 4 min of withdrawal[167]. The fall in cAMP levels reflects mainly the rate of decay of adenyl cyclase activity to its resting state; the rate of cAMP hydrolysis by phosphodiesterase is not affected by glucagon administration or withdrawal. Hence, glucagon–receptor complex formation appears to be rapid and reversible.

In the intact cell, then, the receptor is specific for a hormone, responds to the hormone over a well-defined concentration range, and is rapidly and reversibly activated by the hormone. These properties should persist in cell-free preparations.

(b) Characteristics of adenyl cyclase activation in cell-free systems

(i) Specificity and dose–response relationships

Addition of glucagon to isolated-liver plasma membranes results in an 8–10 fold stimulation of adenyl cyclase activity. This effect is specific for native glucagon, since secretin, and a great number of other peptide and protein hormones, including the biologically inactive glucagon derivative DH-glucagon, were found to be ineffective stimulators of adenyl cyclase activity in isolated membranes[175,176]. In close agreement with liver perfusion studies, half-maximal activation of adenyl cyclase in isolated membranes is obtained at about 10^{-9} M, varying somewhat with the incubation conditions. Thus, while Pohl, *et al.*[175] found an apparent activation constant (K_a) for glucagon of $4 \cdot 10^{-9}$ M, Lin *et al.*[163] have recently reported an apparent K_a of about $7 \cdot 10^{-10}$ M, obtained by using the synthetic substrate AMP–PNP, GTP and omitting chelating agents from the incubation medium, factors shown to modify hormonal stimulation in isolated adenyl cyclase systems. DH-glucagon is ineffective in stimulating the liver adenyl cyclase system; however, it is a potent *inhibitor* of glucagon action[176] with an apparent dissociation constant of about 2 to $4 \cdot 10^{-8}$ M. This constant is almost identical to the apparent K_a of glucagon obtained under identical conditions. DH-glucagon was found to inhibit glucagon activation of adenyl cyclase not only in liver, but also in fat cells and a β-cell tumor of the Syrian hamster. These results indicate that even though the discriminatory function of the receptor is little affected by the absence of histidine, initiation of action is absolutely dependent on its presence; this characteristic being similar in several tissues[176,104]. The dose–response relations for glucagon action in the intact liver cells and for glucagon action on adenyl cyclase in isolated liver membranes have therefore been shown to correlate well. Similarly, epinephrine was found to elevate cAMP levels in the rat heart[88] over the same concentration range over which it stimulated adenyl cyclase activity in heart homogenates[86].

These results illustrate that in some instances hormone–receptor function is completely preserved upon partial isolation. In other instances, however, a distinct divergence of the dose–response curves obtained with intact cells on cAMP levels and with isolated membranes on adenyl cyclase activity was found. For example, while ACTH and epinephrine affect cAMP levels in isolated rat-fat cells with an apparent K_a of $5 \cdot 10^{-10}$ M, and $5 \cdot 10^{-8}$ M respectively[106], in isolated membrane preparations these hormones stimulate adenyl cyclase half-maximally at $3 \cdot 10^{-7}$, and $6.0 \cdot 10^{-6}$ respectively[107].

Several possibilities may account for these discrepancies: (1) the membrane system may have been damaged during isolation, so that it no longer reflects the physiologic state that predominates in the intact cell. Since phospholipids have been demonstrated to play a specific role in the coupling process, it is likely that such damage might be due to alterations (oxidation?) of membrane phospholipids. (2) The intact cell system may have a hormone-concentrating device that locally increases hormone concentration and which has become inoperative in the isolated membrane. Experimentally, this device should consist of hormone-specific binding sites of high affinity present in large amounts on the cell surface. The advantages for the cell of having such a mechanism is that it would aid in assuring that hormones, which circulate at very low levels in blood, "find" the extremely low number of receptors present on the surface. (3) Non-linear coupling processes may intervene between hormone–receptor complex formation and stimulation of adenyl cyclase. (4) The coupling process may be under the regulatory control of soluble metabolites lost or washed out during preparation of adenyl cyclase containing membranes. This possibility is suggested by the recent discovery that purine nucleosides and nucleotides modify hormonal stimulation of adenyl cyclase systems[177–180] (see also below). (5) Finally, the shift of the dose–response relationship in broken-cell preparations towards higher concentrations may be due either to (a) an incubation artifact such as presence of membrane-

associated hormone-degrading mechanisms whose effect would be notice-
able at sub-saturating hormone concentrations, or (b) to erroneous experimental design such as not allowing enough time for the receptor-binding
reaction to reach equilibrium before adenyl cyclase activity resulting from
hormone-receptor interaction is determined. Numerous hormone-binding
studies have shown that it may take as long as 15–20 min at 30–37° to
achieve maximal levels of binding[181–186].

(ii) Kinetic aspects of adenyl cyclase activation

Looking at data from liver-perfusion experiments, stimulation of the
liver-membrane adenyl cyclase system by glucagon is rapid, as long as
saturating concentrations of the hormone are used. The activation rate at
submaximal concentrations, however, depends on the incubation conditions.
With high concentrations of ATP, the presence of EDTA and an ATP-
regenerating system, there is no significant lag period between hormone
addition and attainment of the stimulated state (Fig. 3). However, by

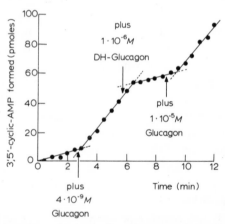

Fig. 3. Time course of cAMP production by liver-plasma membranes before and after succes-
sive addition of $4 \cdot 10^{-9}$ M glucagon, $1 \cdot 10^{-6}$ M des-1-histidine glucagon (DH-glucagon)
and $1 \cdot 10^{-5}$ M glucagon. Liver membranes were incubated for adenyl cyclase activity in a
final volume of 1.3 ml in a 13 × 100 mm test tube at 30°. Final composition of the incubation
medium was: 3.2 mM [α-^{32}P]ATP (approximately 65 cpm per pmole), 5 mM MgCl$_2$, 1.0
mM cAMP, 1.0 mM EDTA, 20 mM creatine phosphate, 1 mg per ml creatine kinase, and
25 mM Tris-HCl, pH 7.5. Test tube and reagents were temperature equilibrated before
initiation of the reaction. Formation of cAMP was followed by removing 50-μl aliquots and
terminating the reaction by adding them to 100 μl of 40 mM ATP, 12.5 mM [^{3}H]cAMP
(approximately 10000 cpm) and 1% sodium dodecyl sulphate followed by immediate boiling
for 3.5 min. [^{32}P]cAMP formed from [α-^{32}P]ATP was determined by the method of Krishna
et al. (1968). (From ref. 164)

modifying the incubation conditions (0.1 mM AMP–PNP, no EDTA, no ATP-regenerating system or GTP) Lin et al.[163] showed a lag period varying from about 30 sec at 10^{-7} M to as much as 4 min at $5 \cdot 10^{-10}$ M. Addition of GTP significantly reduced this lag time, especially at the low glucagon concentration, and simultaneously increased the magnitude of activation.

This lag period of adenyl cyclase is not unique to liver tissue. Bockaert et al., have recently reported slow activation of an adenyl cyclase system from epithelial cells of frog bladder with low concentrations of oxytocin[186]. We have observed a similar phenomenon working with an arginine–vasopressin–(AVP)-sensitive adenyl cyclase system from beef renal medulla when tested at 10^{-10} M and 10^{-9} M AVP (Fig. 4). As yet no definite judgement can be made as to the physiologic meaning of the lag periods

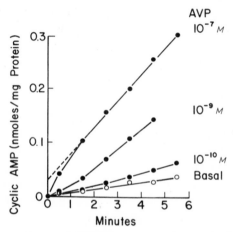

Fig. 4. Time course of cAMP production by beef renal medullary membranes and various concentrations of 8-arginine vasopressin (AVP). Beef renal medullary membranes, collected from the 37/41.5% sucrose interface of a discontinuous sucrose density gradient (from bottom to top): 10 ml 41.5% sucrose, 7.5 ml 37% sucrose, 5 ml 32.5% sucrose and 15 ml of a suspension of $1500 \times g$ particles prepared from a homogenate of beef renal medulla) after a 2-h centrifugation at 27000 rpm in a Beckman SW 27 rotor, were incubated at 37° in a final volume of 0.05 ml of medium containing 0.08 mM [α-^{32}P]ATP (2500 cpm per pmole), 2.0 mM MgCl$_2$, 1.4 mM EDTA, 1.0 mM cAMP, 20 mM creatine phosphate, 0.2 mg per ml of creatine kinase, 25 mM bis–tris–propane–HCl buffer (Sigma) pH 8.0, and the indicated concentrations of AVP. Test tubes containing all the ingredients for the assay were "pre-incubated" for 1 min at 37° prior to the addition of membranes (20-μl aliquots) with which the reaction was started. At the indicated times the reaction was stopped by addition of 100 μl of medium containing 40 mM ATP, 10 mM [^{3}H]cAMP (approximately 10000 cpm) and 1% sodium dodecyl sulfate followed by boiling for 3.5 min. [^{32}P]cAMP formed from [α-^{32}P]-ATP was determined by the method of Krishna et al.[164].

observed at low hormonal concentrations. In the isolated liver-plasma membranes it seems that the combination of glucagon concentration and availability of nucleotides will be decisive with respect to the rapidity of onset of activation of adenyl cyclase systems.

(iii) Reversibility of hormone–receptor interaction

Work carried out with the glucagon-sensitive adenyl cyclase system from rat-liver plasma membranes demonstrated unequivocally that activation of adenyl cyclase by hormones is a reversible process[187]. Thus, it was shown that: (*1*) addition of excess of DH-glucagon to a glucagon-stimulated adenyl cyclase system proceeding at its half-maximal rate results, within one minute, in the return to basal levels of activity (Fig. 3); (*2*) removal of glucagon by washing results in loss of the activated state of the enzyme, provided the concentrations of the peptide hormone to which the membranes were first exposed did not exceed the apparent K_a by more than ten times[188]; (*3*) dilution of the incubation medium in which a glucagon-stimulated adenyl cyclase is proceeding at a given submaximal rate with glucagon-free incubation medium results in immediate and proportional loss of the stimulated rate of cAMP production; and (*4*) submaximally stimulated glucagon activities decay as a function of incubation time, i.e., they are curvilinear, possibly due to significant inactivation of glucagon by a glucagon-specific degrading system present in liver plasma membranes[181,189]. These studies indicate that adenyl cyclases are reversibly activated. The proportion of the enzyme in the stimulated state depends on the instantaneous concentration of free hormone in the medium, and by inference, on the proportion of hormone–receptor complex.

With less purified membrane systems (fat-cell ghosts and ACTH, low-speed particles from rat liver and glucagon) higher hormone concentrations resulted in persistent stimulation of adenyl cyclase activity which could not be reversed by simple washing procedures[149,190]. A similar result was obtained with purified liver-plasma membranes when high concentrations of glucagon (approx. 500 times apparent K_a) were used[188]. It is likely that the observed persistent effects are due to binding of the respective hormones to non-specific sites on the membrane preparations, which upon subsequent incubation, release saturating amounts of the hormone.

The following may therefore be concluded about cell-free studies: (*1*) the characteristics of adenyl cyclase systems in the isolated state may, in some instances, be very close if not identical to the characteristics of these systems in the intact cell; (2) the isolated systems may contain functional receptors with respect to affinity and specificity and to initiation of hormone action and (3) studies on the mode of action of receptors in the isolated system will lead us to an understanding of the primary events in hormone action, provided the particular system under study has been validated with respect to hormone specificity, dose–response relation *and* kinetics.

(c) Hormone binding; relation to adenyl cyclase activation

Rodbell and collaborators[175,176,181,187,189,191,192] carried out an extensive study of the properties of binding of glucagon to specific binding sites in rat-liver plasma membranes and correlated their results with the characteristics of the receptor-dependent stimulation of adenyl cyclase activity.

Binding of biologically active iodinated glucagon to liver plasma membranes has two main characteristics: (1) it is specific for glucagon or DH-glucagon; no other peptide or protein hormone tested interferes with glucagon and (2) binding occurs over the same concentration range over which glucagon activates the adenyl cyclase system.

With respect to rapidity and reversibility, however, the binding appears to be complex and its relation to adenyl cyclase activation is not clear. First, binding is slow, taking about 10–15 min to reach constant values at $4 \cdot 10^{-9}$ M glucagon. Second, as the binding reaction proceeds it is associated with a progressive loss of reversibility: addition of chase quantities of unlabeled glucagon 1 min after intitiation of the binding reaction resulted in total release (dissociation) of labeled glucagon from the membranes in 1 or 2 min. But, a glucagon chase at 5 and 15 min results in rapid release of only about 30 and 10% of the label respectively, the rest being bound almost irreversibly. Third, this time-dependent loss of reversibility of binding is overcome by the addition of micromolar concentrations of GTP and GDP or of millimolar concentrations of ATP. Thus, in the presence of 10^{-6} M GTP, addition of a glucagon chase after 15 min leads to total dissociation of the label within a few minutes. Finally, even in the presence of GTP or ATP to assure reversible binding, the binding reaction does not seem to proceed fast enough to account for the rapid activation of the receptor-coupled adenyl cyclase system. For example, it was found that in the presence of 1 mM EDTA, 3.2 mM ATP, and ATP-regenerating system, stimulation of the adenyl cyclase by $4 \cdot 10^{-9}$ M glucagon is achieved within 10–20 sec, but occupation of the glucagon-specific binding sites is only 10–20% of that achieved with prolonged incubation. In addition, DH-glucagon, added 5 min after glucagon, completely inhibited hormonal stimulation within 1 min, while displacing only 10–20% of glucagon from its binding sites. This cannot be reconciled with the receptor-occupation theory of hormone action according to which adenyl cyclase activation should be proportional to hormone-occupied binding sites.

Such results seem to indicate that a large proportion of the glucagon specific sites (at least 80–90%) does not participate in the activation of the adenyl cyclase system and might therefore be related to degrading or concentration functions. Specific inactivation of glucagon, unrelated to the mechanism of adenyl cyclase activation, has been described in liver membranes. Correlative phenomena will have to be carefully considered before non-activating hormone-binding can be interpreted. For example, there are indications of interaction (cooperativity) between glucagon-specific binding sites, suggested by the finding that more labeled glucagon dissociates in the presence of a glucagon chase than in the presence of a saturating DH-glucagon chase[188,193]. Furthermore, results obtained in hormone-sensitive adenyl cyclase systems from several tissues (*e.g.*, rat liver, beef renal medulla and cat heart), indicate that translation of the coupling signal, generated by hormone–receptor interaction, into enzyme activation may have several slow components.

Examples of slow components in adenyl cyclase activation were found both in the liver and the renal system. Thus, in the liver-membrane system with low glucagon concentrations, a considerable lag between initiation of binding and adenyl cyclase activation may exist if the system is tested under nucleotide-limiting conditions without chelating agents[188]. Stimulation of the beef renal medullary membrane adenyl cyclase by 10^{-7} M AVP is associated with a "burst" phase (initial velocity greater than steady-state velocity), indicating a transient state of activity that is more active than the steady state of activity (Fig. 4). Such burst phases, although not specifically commented on, have been reported in the data for fluoride-stimulated activity in rat testicular adenyl cyclase preparations by Murad *et al.*[126], and for basal adenyl cyclase activity in membranes of a β-cell tumor by Goldfine *et al.*[104]. This transient state of activity may be a general phenomenon which has not yet received enough attention. The existence of lag and burst phases clearly preclude any simple correlations between the time-course of hormone binding and the time-course of hormone-dependent enzyme activation. Activation must be the resultant of a slow (lag-generating), hormone-binding function and a complex, unexplained burst-generating enzyme-activation function. It is conceivable that at a given hormone concentration and in the presence of substances such as nucleotides and/or chelating agents, these two phenomena "cancel" themselves yielding "linear" time curves.

In conclusion, hormone-binding studies are just beginning and are not always comprehensible. In the one study[193] that compared binding to function, specificity and apparent affinity parameters correlated well with that required for receptor activity. However, it appears that a large proportion of specific binding, perhaps as much as 80–90%, is not related to adenyl cyclase activation.

The significance of hormone-binding which is apparently not coupled to adenyl cyclase activation has yet to be explained.

(d) Factors affecting coupling of hormone–receptor interaction to adenyl cyclase activation

(i) Role of phospholipids in coupling

Hormone-sensitive adenyl cyclase systems are firmly bound to plasma membranes of cells. Many of the properties of these complex multimolecular systems are, therefore, conditioned by the environment in which they are located. Rodbell *et al.* found that treatment of fat-cell ghosts with phospholipase C or of liver-plasma membranes with high concentrations of phospholipase A resulted in total loss of enzymatic activity. However, treatment with lower concentrations of phospholipase A resulted in a selective loss of the responsiveness of the liver system to glucagon, with either unimpaired or stimulated fluoride-sensitive activity. Selective loss of hormonal stimulation also resulted when liver membranes were treated with digitonin[108,192]. Glucagon stimulation could be partially restored by the addition of membrane phospholipids or pure phosphatidyl choline, phosphatidyl ethanolamine or phosphatidyl serine, the most effective[192]. These results suggested that adenyl cyclases, in addition to being dependent on membrane integrity, are dependent on phospholipids. This was the first indirect evidence of a selective role of phospholipids in hormonal stimulation of adenyl cyclase systems.

Levey, working with cat-heart adenyl cyclase systems provided further evidence for a specific role of phospholipids in hormonal response. In a series of elegant experiments[37,194,87], he demonstrated that treatment of heart adenyl cyclase with the non-ionic detergent Lubrol-PX results in loss of the enzyme's response to norepinephrine and glucagon, and that specific phospholipids restore response to either one or the other hormone. Thus, addition of phosphatidyl inositol selectively restored the response to the catecholamine (Fig. 5), and addition of phosphatidyl serine restored response to glucagon.

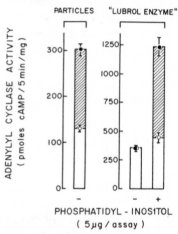

Fig. 5. Effect of phosphatidylinositol on response of Lubrol-solubilized and DEAE-cellulose chromatography purified cat-heart adenyl cyclase to norepinephrine. Open bars represent adenyl cyclase activity determined in the absence of norepinephrine; hatched bars indicate change of activity due to addition of 50 μM norepinephrine. For further experimental details see ref. 87. (Adapted from Levey[87])

(ii) The role of nucleotides and nucleosides in the coupling process

Studies on glucagon binding led Rodbell and collaborators to study the glucagon-stimulated adenyl cyclase using the synthetic substrate AMP–PNP[177,195], or very low concentrations of ATP[187] that do not affect the reversibility of binding. It was found that under either one of these conditions glucagon-stimulated adenyl cyclase activity is strongly dependent on the addition of low concentrations (10^{-8} to 10^{-6} M) of GTP or relatively high concentrations (10^{-5} to 10^{-3} M) of ATP (Fig. 6). The concentration range over which these two nucleotides promote glucagon stimulation is the same as that over which they promote reversibility of binding. The extremely low concentration of GTP that affected this system suggested that the guanyl nucleotide and not the adenyl nucleotide is the

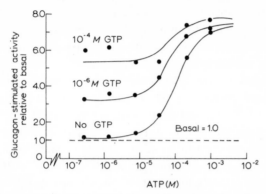

Fig. 6. Effect of ATP and GTP on stimulation of liver-plasma membrane adenyl cyclase activity by 5 μM glucagon. Activities in the presence of glucagon were determined in the absence and the presence of 10^{-6} M and 10^{-4} M GTP. Activities relative to basal were calculated by dividing the activity obtained in the presence of glucagon at each ATP concentration by the respective activity obtained in the absence of glucagon. Incubations were for 2 min at 30°. Rest of incubation conditions are described in legend to Fig. 3.

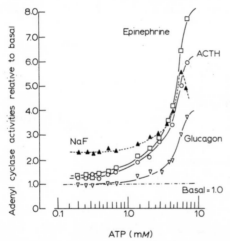

Fig. 7. Effect of ATP on stimulation of fat-cell "ghost" adenyl cyclase activity by ACTH, glucagon, epinephrine and NaF. Fat-cell "ghosts" (for details of preparation see ref. 58) were incubated in final volume of 0.05 ml for 10 min at 30° with the indicated concentrations of ATP in the absence (Basal) and the presence of 10 μg per ml ACTH (ACTH), 10 μg per ml glucagon (Glucagon), 10 μg per ml epinephrine (Epi) and 10 mM NaF (NaF). Rest of incubation conditions were: 5 mM $MgCl_2$, 10 mM theophylline, 0.05% albumin, 10 mM creatine phosphate, 0.2 mM creatine phosphate, and 25 mM Tris–HCl, pH 7.5. At each ATP concentration the stimulation of cAMP production was expressed relative to that obtained under basal conditions.

References p. 101

natural effector. GMP–PCP, a non-phosphorylating analogue of GTP was also effective in enhancing glucagon stimulation[177]. It seems therefore, that the effect is probably due to an interaction of the ligand and regulatory site as opposed to a chemical modification (phosphorylation) of one of the components of the adenyl cyclase system.

Purine nucleoside triphosphates were shown to enhance hormonal stimulation in fat-cell ghosts (Fig. 7), pancreatic β-cells (Fig. 8 and ref. 104), human platelets[178], and beef renal medulla (see below).

Birnbaumer partially purified membrane particles from beef-kidney medulla that are enriched in an adenyl cyclase system highly sensitive to neurohypophyseal hormones (apparent K_a for AVP, $1 \cdot 10^{-9}$ M), and initiated studies exploring the effects of nucleotides on the action of AVP in this system. As illustrated in Fig. 9, ATP has both stimulatory and inhibitory effects on AVP response. The stimulatory effect is seen between 5 and 90 μM ATP; the inhibitory effect becomes evident when the concentration of ATP is increased further to 1.0 mM.

Addition of GTP results in inhibition of AVP-stimulated activity without affecting basal activity. Concentrations of GTP as low as $5 \cdot 10^{-8}$ M clearly

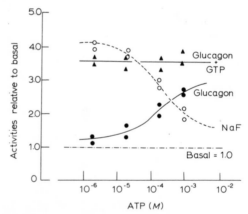

Fig. 8. Effect of ATP and GTP on response of adenyl cyclase activity of pancreatic β-cells to glucagon and fluoride. $1000 \times g$ particles prepared from insulin-secreting tumors of the Syrian (golden) hamster (for details see ref. 104) were incubated in 0.06 ml for 2.5 min at $30°$ with the indicated concentrations of ATP in the presence of no addition (Basal), 10 μg per ml glucagon (Glucagon), 10 μg per ml gluagon plus 10^{-5} M GTP (Glucagon plus GTP) and 10 mM NaF (NaF). Rest of incubation conditions were: 25 mM creatine phosphate, 1 mg per ml creatine kinase, 0.33% albumin (human), 5 mM MgCl$_2$, and 50 mM Tris–HCl, pH 7.8. At each ATP concentration the stimulation of cAMP production was expressed relative to that obtained under basal conditions. (From ref. 104)

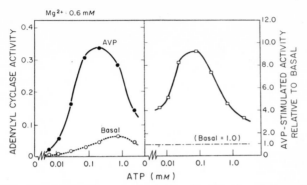

Fig. 9. Effect of ATP concentration on stimulation of beef-renal medullary adenyl cyclase activity by 10^{-7} M AVP. AVP-stimulated activity relative to basal was calculated by dividing the activity obtained in the presence of AVP at each ATP concentration by the respective activity obtained in the absence of AVP. Incubations were for 10 min. Rest of conditions are described in legend to Fig. 4. Adenyl cyclase activities are expressed as nmoles cAMP formed per mg protein per 10 min. (From ref. 200)

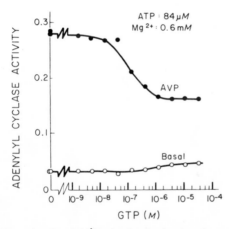

Fig. 10. Effect of GTP on basal and AVP-stimulated adenyl cyclase activities in beef-renal medullary membranes. Renal medullary membranes were incubated in the presence of 84 μM ATP and 0.6 mM Mg^{2+} (in excess over 1.4 mM EDTA), and the indicated concentrations of GTP. AVP when present was 10^{-7} M. Rest of incubation conditions are described in legend to Fig. 4. Adenyl cyclase activities are expressed as nmoles cAMP formed per mg protein per 10 min. Data from Birnbaumer[200].

affect this system; half-maximal inhibition is obtained with about $2 \cdot 10^{-7}$ M (Fig. 10). Thus the GTP-dependent step discovered in the glucagon-sensitive adenyl cyclase system in liver-plasma membranes does not seem to be an invariant feature of hormone-sensitive adenyl cyclase systems.

References p. 101

TABLE VI

Effects of arginine–vasopressin (AVP) and prostaglandins (PG) on adenyl cyclase activity in beef renal medullary membranes

Additions	Adenyl cyclase activity	Change due to addition of	
		PGE_1	AVP
None	0.068 ± 0.008		
$PGF_{2\alpha}$	0.062 ± 0.005		
PGE_1	0.176 ± 0.009	0.108	
AVP	0.285 ± 0.002		0.217
$PGF_{2\alpha} + AVP$	0.273 ± 0.004		
$PGE_1 + AVP$	0.273 ± 0.001	0.106	0.212

Beef-renal medullary membranes (37.0/41.5 interface) were incubated as described in labeling to Fig. 4. Final concentration of ATP was 1.0 mM. When present AVP was $1 \cdot 10^{-7}$ M and prostaglandins were 10 μg/ml.

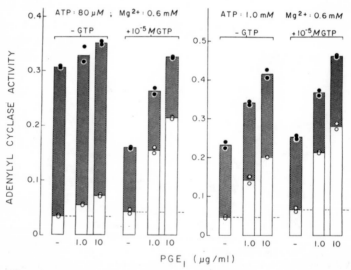

PGE$_1$ (μg/ml)

Fig. 11. Effect of GTP and ATP on stimulation of renal medullary adenyl cyclases by PGE$_1$ and AVP. Beef-renal medullary membranes (37/41.5% sucrose interface) were incubated for 10 min at 37° with the indicated concentrations of ATP, Mg^{2+} (excess over 1.4 mM EDTA), GTP and prostaglandin E$_1$ (PGE$_1$). Rest of incubation conditions are described in the legend to Fig. 4. Open bars represent adenyl cyclase activity determined in the absence of AVP; hatched bars represent change in activity due to addition of 10^{-7} M AVP. Adenyl cyclase activities are expressed as nmoles cAMP formed per mg protein per 10 min. Data from Birnbaumer[200].

This conclusion is supported by the findings of Wolff and Cook[180] who determined that the TSH response in beef-thyroid membranes is preferentially stimulated by ITP rather than GTP, although the latter was also active.

Beef-renal medullary membranes also contain a prostaglandin(PG)-sensitive adenyl cyclase system that is distinct from the AVP-sensitive system, since changes in activities due to saturating concentrations of AVP and PG are additive (Table VI). The effect of PGE_1 is dependent on the addition of micromolar concentrations of GTP or millimolar concentrations of ATP. Fig. 11 illustrates both effects of GTP, that inhibiting AVP response and that stimulating PGE_1 response. It also shows that the effects of ATP and GTP on either adenyl cyclase system are not additive, suggesting that ATP, at concentrations approaching 1.0 mM, mimics the actions of GTP on both renal medullary adenyl cyclases.

The characteristics of the stimulatory effect of ATP on AVP response seen between 5 and 90 μM ATP, were further investigated. It was found that this effect is not specific for ATP. Adenosine and AMP, but *not* cAMP, cGMP, GMP, or cIMP, mimic the effect of ATP in maximally stimulating the response of the renal system to AVP. The effects of adenosine and

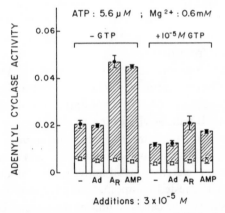

Fig. 12. Effect of adenine, adenosine, and AMP on basal, and AVP-stimulated adenyl cyclase activities in beef-renal medullary plasma membranes. Renal medullary membranes (37/41.5% sucrose interface) were incubated for 10 min at 37° at the indicated concentrations of ATP, Mg^{2+} (excess over 1.4 mM EDTA), GTP, adenine (Ad), adenosine (A_R), and AMP. Open bars represent activities determined in the absence of AVP and hatched bars represent the change in activity due to addition of 10^{-7} M AVP. Rest of incubation conditions are described in the legend to Fig. 4. Adenyl cyclase ctivities are expressed as nmoles cAMP formed per mg protein per 10 min. (From ref. 200)

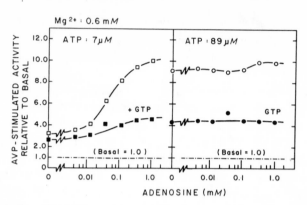

Fig. 13. Effect of varying concentrations of adenosine on hormonal stimulation of renal medullary membrane adenyl cyclase activity determined at suboptimal and at optimal ATP concentrations. Incubations were for 10 min at 37°. When present GTP was 10^{-5} M and AVP, 10^{-7} M. Rest of incubation conditions are described in the legend to Fig. 4. (From ref. 200)

AMP at $3 \cdot 10^{-5}$ M on the AVP-stimulated activity determined at 5 μM ATP are shown in Fig. 12. It was also found that the stimulatory effects of ATP and adenosine are not additive (Fig. 13), suggesting that these compounds act at the same site or through the same process. Neither adenosine (Fig. 13) nor AMP (not shown) interfere with the inhibitory effect of GTP on AVP stimulation of adenyl cyclase, suggesting that adenosine or AMP act at sites different from those of GTP.

The stimulation of hormone action by adenosine and AMP is reminiscent of effects seen in 1970 by Sattin and Rall[196] in brain slices. These purine derivatives led to increased cAMP levels and potentiated the stimulatory action of histamine. It will be interesting to determine whether these two phenomena are related from a mechanistic point of view.

Effects of nucleotides and nucleosides were analyzed by expressing hormonal stimulation relative to basal activity. This mode of analysis corrects for any effect the tested compound may have on basal activity, i.e., on the functional state of the catalytic unit of the system. Depending on the system studied, purine derivatives (GTP, adenosine, AMP and ITP) may have profound effects on the catalytic unit of the system, with or without affecting the susceptibility of the system to hormones. An example of this was recently reported by Fain et al.[197], who found that adenine nucleosides, particularly 2-deoxyadenosine, at concentrations about 9.1 mM inhibit fat-cell adenyl cyclase activity without affecting the ratio between the

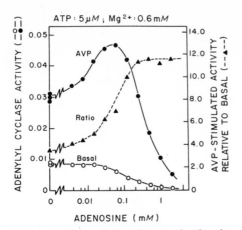

Fig. 14. Effect of varying concentrations of adenosine on adenyl cyclase activity determined in the absence and the presence of 10^{-7} M AVP. Concentrations of ATP and Mg^{2+} (excess over EDTA) were those indicated on the figure. Incubations were for 10 min at 37°. Rest of incubation conditions are described in the legend to Fig. 4. Adenyl cyclase activities are expressed as nmoles cAMP formed per mg protein per 10 min. (From ref. 200)

remaining basal and hormonally-stimulated activities. Studies by Leray et al.[179], on the other hand, demonstrated that in liver of adrenalectomized rats, addition of GTP results in increase of both epinephrine stimulation and basal activity. Fortuitously, GTP exerts little, if any, action on basal activity of the renal medullary membrane adenyl cyclase and thus facilitates analysis and interpretation; but this is not so with adenosine. At 0.5 mM, for example, this nucleoside both stimulated AVP response and strongly inhibited the catalytic process (see Fig. 14). Final interpretation of effects of nucleosides and nucleotides will have to take these "side-effects" into account.

In conclusion, experiments with beef-renal medullary membranes suggest that ATP can interact with three distinct sites of the AVP-sensitive adenyl cyclase system: (1) catalytic sites, serving as a substrate, (2) regulatory site I, enhancing AVP response (this effect appears to be mimicked by adenosine and AMP), and (3) regulatory site II, inhibiting AVP response (mimicking the inhibitory effect of GTP). Like the existence of an allosteric site for Mg^{2+} in other adenyl cyclase systems, the actual existence of multiple regulatory sites in the AVP-sensitive adenyl cyclase system is still a matter of speculation. The lesson to be learned from these studies, however, is that nucleotides and nucleosides, up to now purine derivatives, regulate

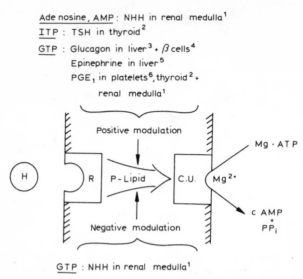

Adenosine, AMP : NHH in renal medulla[1]
ITP : TSH in thyroid[2]
GTP : Glucagon in liver[3] + β cells[4]
        Epinephrine in liver[5]
        PGE₁ in platelets[6], thyroid[2] +
        renal medulla[1]

GTP : NHH in renal medulla[1]

Fig. 15. Current pattern of modulation of hormonal stimulation of adenyl cyclase systems by nucleotides and nucleosides. Listed are the purine derivatives which have thus far been shown to affect hormonal stimulation. The abbreviations used are: H, hormone; R, receptor; C.U., catalytic unit; NHH, neuro-hypophyseal hormone; TSH, thyroid-stimulating hormone; PGE₁, prostaglandin E₁. Receptor and catalytic unit are pictured as being located at the outer and the inner surface of the plasma membrane respectively. The coupling process intervening between receptor–hormone interaction and stimulation of the catalytic unit is pictured by the arrow connecting receptor and catalytic unit. It is assumed that lipids and purine derivatives play critical roles in modulating the coupling process: *1* from Birnbaumer[200], *2* from Wolf and Cook[180], *3* from Rodbell et al.[55], *4* from Goldfine et al.[104], *5* from Leray et al.[179] and *6* from Krishna et al.[178].

the expression of hormonal stimulation by acting, either as positive or negative modulators, depending on the system studied. A schematic overview of our current knowledge of factors affecting modulation of hormonal response in adenyl cyclases is shown in Fig. 15.

7. Mode of activation of adenyl cyclase. Role of magnesium ion

Studies by Birnbaumer et al.[58] revealed that one of the kinetic parameters affected by ACTH and fluoride ion in the rat fat-cell ghost adenyl cyclase system is the apparent affinity of Mg^{2+} for a site distinct from the catalytic site at which $Mg \cdot ATP$ is converted to cAMP and PP_i. The following evidence for the above conclusion has been obtained: (*1*) the affinity of the catalytic site for $Mg \cdot ATP$ is unaltered by hormone or fluoride; (*2*)

Hill plots indicate that the order of the reaction of Mg^{2+} with adenyl cyclase is 2; and (3) the apparent affinity of the system for Mg^{2+} is markedly enhanced in the presence of ACTH and fluoride. Since Mg^{2+}, far in excess of that required for the formation of $Mg \cdot ATP$, enhances basal activity to levels comparable to those seen in the presence of hormone or fluoride at lower Mg^{2+} concentrations, it was also concluded that the activity of the catalytic site is influenced by an extra-catalytic Mg^{2+} site. Moreover, the presence of the stimulatory agents results in an altered behavior of the allosteric regulatory Mg^{2+} site. There is suggestive evidence for the existence of an allosteric site for Mg^{2+} in at least two other systems. Drummond and Duncan[59] have shown in cardiac tissue that Ca^{2+}, which at millimolar concentrations inhibits enzymatic activity, appears to do so by competing with Mg^{2+} at a site distinct from the catalytic site. In brain and heart, Perkins and collaborators[198,199] have found that preliminary exposure of adenyl cyclase-containing membranes to fluoride ion results not only in increased V_{max} of the catalytic process but also in decreased apparent K_a for Mg^{2+}. It seems therefore possible that the coupling process through which hormones and fluoride stimulate adenyl cyclase activity, shown both in fat[58] and liver-plasma membranes[64] to be separate, do so in part by regulating the affinity of an allosteric site for Mg^{2+}.

Stimulation of adenyl cyclase activity is not always associated with an apparent increase in affinity for Mg^{2+}. For example, in Drummond and Duncan's experiments[59], stimulation of adenyl cyclase activity seemed to be due exclusively to increased V_{max}. In studies by Pohl et al.[175], the Mg^{2+} dependence of the liver-plasma membrane adenyl cyclase system was of such complexity that it was impossible to ascertain the existence of an allosteric site for Mg^{2+}. Thus, the actual existence of a second allosteric site for Mg^{2+}, although attractive, is still a matter of speculation. Recent curve-fitting experiments carried out by Dr. Christoph Dehaen (personal communication) demonstrate that the experimental data obtained with fat-cell ghosts and cardiac tissue particles fit equally well a model in which no allosteric site for Mg^{2+} exists.

In this model the enzyme's catalytic site has extremely high affinity for free ATP which inhibits catalysis. Stimulation of the catalytic activity could thus result either through reduction of affinity for free ATP or through increased catalytic efficacy (V_{max}). In this model, full activity requires high concentrations of Mg^{2+} (to complex the inhibitory free ATP). Accordingly, adjustment to the fat-ghost system requires the change in affinity for free

ATP to predominate over the change of V_{max}, and adjustment to the myocardial system requires the change of V_{max} to predominate over the change of affinity.

It should be mentioned here that other divalent cations may also play a role in activation of the catalytic unit of the cyclase system. Thus, calcium has been shown to be necessary in trace (submicromolar) concentrations for both basal and arginine–vasopressin stimulated activities in porcine renal medullary membranes[183]. Other systems show increased activity in the presence of EDTA or EGTA[175], suggesting that trace metals may also be involved.

The recent discovery of regulatory effect of purine nucleosides and nucleotides on expression of hormonal stimulation in various adenyl cyclase systems raises the necessity of exploring the effects of divalent cations on this facet of adenyl cyclase regulation.

8. Final remarks

It should be evident that considerable information has been accumulated in recent years concerning the molecular nature and function of hormone receptors coupled to adenyl cyclases. Probing the kinetics of adenyl cyclase activation has provided new insight into the mode of action of hormones. However, the kinetics of the hormonally-induced activation process, the coupling of receptor to adenyl cyclase and the architectonic organization of the adenyl cyclase system in the plasma membrane are basic problems still unresolved. We may envisage an uneven distribution of hormone receptors and adenyl cyclases organized in clusters across the membrane surface, which might account for a more complex non-Michaelian kinetics. The mode of coupling of the hormone receptors and the adenyl cyclases is an enigma. We may conceive two possibilities: (1) either the receptor and the catalytic unit are permanently coupled, or (2) hormone receptors 'float' within the membrane independent of the catalytic unit. In the latter case, coupling and concomitant stimulation of activity is established only while hormone binding to receptors take place. Temporary coupling of the hormone receptor with the catalytic unit could be the basis for time delays and bursts of activity which are observed in some of the adenyl cyclase systems studied.

Finally, the molecular entities of this complex system have yet to be isolated and characterized. Attempts at solubilization and separation of the molecular components are in their initial stages. These are formidable tasks since there is the possibility of losing hormonal responsiveness at each step of the separation process. Nevertheless, much has been recently learned concerning the various factors involved in the coupling of hormone–receptor reaction to adenyl cyclase activation. It is our opinion and our hope that these efforts will soon lead to the isolation of the receptor and catalytic component of the adenyl cyclase system.

ACKNOWLEDGEMENTS

It is a pleasure to thank Dr. Sol Sepsenwol for the language revision. The authors acknowledge support by grants No. RF 71074 from the Rockefeller Foundation, Project Grant No. R 01 HD 06513-01 and Program Grant No. P 01 HD 06273-01 from the U. S. Public Health Service.

REFERENCES

1   E. W. Sutherland, I. Øye and R. W. Butcher, *Recent Progr. Hormone Res.*, 21 (1965) 623.
2   E. W. Sutherland, G. A. Robinson and R. W. Butcher, *Circulation*, 37 (1968) 37.
3   G. A. Robison, R. W. Butcher and E. W. Sutherland, *Ann. Rev. Biochem.*, 37 (1968) 149.
4   E. W. Sutherland, *J. Am. Med. Assoc.*, 214 (1970) 1281.
5   G. A. Robison, R. W. Butcher and E. W. Sutherland, *Cyclic AMP*, Academic Press, New York, 1971.
6   E. V. Jensen and E. R. DeSombre, in: L. Martini and V. H. T. James (Eds.), *Current Topics in Experimental Endocrinology*, Vol. 1, Academic Press, New York, 1971, p. 229.
7   E. V. Jensen and E. R. DeSombre, *Ann. Rev. Biochem.*, 41 (1972) 203.
8   J. Gorski, D. Toft, G. Shyamaca, D. Smith and A. Notides, *Recent Progr. Hormone Res.*, 24 (1968) 45.
9   I. S. Edelman and D. D. Fanestil, in: G. Litwack (Ed.), *Biochemical Actions of Hormones*, Vol. 1, Academic Press, New York, 1970, p. 321.
10   S. Liao and S. Fang, *Vitamins Hormones*, 27 (1969) 17.
11   E. E. Baulieu, A. Alberga, I. Jung, M. C. Leveau, Ch. Mercier-Bodard, E. Milgrom, J. P. Raynaud, C. Raynaud-Jammet, H. Rochefort, H. Truong and P. Robel, *Recent Progr. Hormone Res.*, 27 (1971) 351.
12   H. G. Williams-Ashman and A. H. Reddi, *Ann. Rev. Physiol.*, 33 (1971) 31.
13   G. Tomkins and E. S. Maxwell, *Ann. Rev. Biochemistry*, 32 (1963) 677.
14   G. M. Tomkins, D. W. Martin Jr., R. H. Stellwagen, J. D. Baxter, P. Namont and B. B. Levinson, *Cold Spring Harbor Symp. Quant. Biol.*, 35 (1970) 635.
15   B. W. O'Malley, W. L. McGuire, P. O. Kohler and S. G. Korenman, *Recent Progr. Hormone Res.*, 25 (1969) 105.
16   G. E. Mueller, B. Vonderhaar, V. H. Kin and M. Le Mahieu, *Recent Progr. Hormone Res.*, 28 (1972) 1.
17   O. Hechter and T. Braun in: M. Margoulies and F. C. Greenwood (Eds.), *Structure–Activity Relationships of Protein and Polypeptide Hormones*, Excerpta Medica, ICS, Amsterdam, No. 241, 1971, p. 211.
18   J. F. Kuo and P. Greengard, *Proc. Natl. Acad. Sci. (U.S.)*, 64 (1969) 1349.

19   J. F. Kuo and P. Greengard, *J. Biol. Chem.*, 244 (1969) 3417.
20   T. Okabayashi, A. Yoshimoto and M. Ide, *J. Bacteriol.*, 86 (1963) 930.
21   R. S. Makman and E. W. Sutherland, *J. Biol. Chem.*, 240 (1965) 1309.
22   M. Ide, A. Yoshimoto and T. Okabayashi, *J. Bacteriol.*, 94 (1967) 1317.
23   I. Pastan and R. C. Perlman, *Science*, 169 (1970) 339.
24   D. S. Barkley, *Science*, 165 (1969) 1133.
25   T. M. Konijn, Y. Y. Chang and J. T. Bonner, *Nature*, 224 (1969) 1211.
26   T. E. Mansour, E. W. Sutherland, T. W. Rall and E. Bueding, *J. Biol. Chem.*, 235 (1960) 466.
27   T. E. Mansour and J. M. Mansour, *J. Biol. Chem.*, 237 (1962) 629.
28   E. W. Sutherland, T. W. Rall and T. Menon, *J. Biol. Chem.*, 237 (1962) 1220.
29· R. Fujii and R. R. Novales, *Am. Zoologist*, 3 (1969) 453.
30   O. M. Rosen and S. M. Rosen, *Arch. Biochem. Biophys.*, 131 (1969) 449; ibid. 141.
31   R. R. Novales and W. J. Davis, *Am. Zoologist*, 9 (1969) 479.
32   M. E. Nadley and T. M. Goldman, *Am. Zoologist*, 9 (1969) 489.
33   T. P. Dousa, R. Walter, I. L. Schwartz, H. Sands and O. Hechter, in: P. Greengard, R. Paoletti and G. A. Robison (Eds.), *Advances in Cyclic Nucleotide Res.*, Vol. 1, Raven, New York, 1972, p. 121.
34   P. R. Davoren and E. W. Sutherland, *J. Biol. Chem.*, 238 (1963) 3009, 3016.
35   E. W. Sutherland and T. W. Rall, *J. Biol. Chem.*, 232 (1958) 1077.
36   E. W. Sutherland and T. W. Rall, *Pharmacol. Rev.*, 12 (1960) 265.
37   G. S. Levey, *Ann. N. Y. Acad. Sci.*, 185 (1971) 449.
38   N. I. Swislocki and J. Tierney, *Biochemistry*, 12 (1973) 1862.
39   E. J. Neer, *J. Biol. Chem.*, 248 (1973) 3742.
40   M. Hirata and O. Hayaishi, *Biochem. Biophys. Res. Commun.*, 21 (1965) 361; 24 (1965) 360.
41   M. Hirata and O. Hayaishi, *Biochim. Biophys. Acta*, 149 (1967) 1.
42   M. Tao and F. Lipmann, *Proc. Natl. Acad. Sci (U.S.)*, 63 (1969) 96.
43   M. Tao, M. L. Salas and F. Lipmann, *Proc. Natl. Acad. Sci. (U.S.)*, 67 (1970) 408.
44   I. Pastan, R. L. Perlman, M. Emmer, H. E. Varmus, B. DeCrombrugghe, P. P. Chem and J. Parks, *Recent Progr. Hormone Res.*, 27 (1971) 421.
45   O. Hechter, H. P. Bär, M. Matsuba and D. Soifer, *Life Sci.*, 8 (1969) 935.
46   E. DeRobertis, G. R. D. L. Arnaiz, M. Alberici, R. W. Butcher and E. W. Sutherland, *J. Biol. Chem.*, 242 (1967) 3487.
47   D. Soifer and O. Hechter, *Biochim. Biophys. Acta*, 230 (1971) 539.
48   S. Liao, A. H. Lin and J. L. Tymoczko, *Biochim. Biophys. Acta*, 230 (1971) 535.
49   M. L. Entman, G. S. Levey and S. E. Epstein, *Biochem. Biophys. Res. Comm.*, 35 (1969) 728.
50   M. Rabinowitz, L. DeSalles, J. Meisler and L. Lorand, *Biochim. Biophys. Acta*, 97 (1965) 29.
51   K. Seraydarian and W. F. N. M. Mommaerts, *J. Cell. Biol.*, 26 (1965) 641.
52   I. Øye and E. W. Sutherland, *Biochim. Biophys. Acta*, 127 (1966) 347.
53   M. Rodbell, L. Birnbaumer and S. L. Pohl, *J. Biol. Chem.*, 245 (1970) 718.
54   H. P. Bär and O. Hechter, *Biochim. Biophys. Acta*, 192 (1969) 141.
55   M. Rodbell, L. Birnbaumer, S. L. Pohl and H. M. Krans, *J. Biol. Chem.*, 296 (1971) 1877.
56   J. P. Harwood and M. Rodbell, *J. Biol. Chem.*, 248 (1973) 4901.
57   T. W. Rall, E. W. Sutherland and J. Berthet, *J. Biol. Chem.*, 224 (1957) 463.
58   L. Birnbaumer, S. L. Pohl and M. Rodbell, *J. Biol. Chem.*, 244 (1969) 3468.
59   G. I. Drummond and L. Duncan, *J. Biol. Chem.*, 245 (1970) 976.
60   K. M. J. Menon and M. Smith, *Biochemistry*, 10 (1971) 1186.
61   T. Braun, *Arch. Biochem. Biophys.*, submitted for publication.

62 R. H. Williams, S. A. Walsh and J. W. Ensinck, *Proc. Soc. Exptl. Biol. Med.*, 128 (1968) 279.
63 T. W. Rall and E. W. Sutherland, *J. Biol. Chem.*, 232 (1958) 1065.
64 L. Birnbaumer, S. L. Pohl and M. Rodbell, *J. Biol. Chem.*, 246 (1971) 1857.
65 I. Øye and E. W. Sutherland, *Biochim. Biophys. Acta*, 127 (1966) 347.
66 O. M. Rosen and S. M. Rosen, *Biochem. Biophys. Res. Comm.*, 31 (1968) 82.
67 O. M. Rosen and J. Ehrlichman, *Arch. Biochem., Biophys.*, 133 (1969) 171.
68 H. Sheppard and C. Burghart, *Biochem. Pharmacol.*, 18 (1969) 2576.
69 G. A. Robison, A. Arnold and R. C. Hartmann, *Pharmacol. Res. Comm.*, 1 (1969) 325.
70 K. Abe, G. A. Robison, G. W. Liddie, R. W. Butcher, W. E. Nicholson and C. E. Baird, *Endocrinology*, 85 (1969) 674.
71 K. Abe, R. W. Butcher, W. E. Nicholson, C. E. Baird, R. A. Liddle and G. W. Liddle, *Endocrinology*, 84 (1961) 362.
72 D. Malamud, *Biochem. Biophys. Res. Comm.*, 35 (1969) 754.
73 M. Schramm and E. Naim, *J. Biol. Chem.*, 245 (1970) 3225.
74 B. Weiss and E. Costa, *J. Pharmacol. Exptl. Ther.*, 161 (1968) 310.
75 B. Weiss and J. W. Crayton, *Advan. Biochem. Psychopharm.*, 3 (1970) 217.
76 B. Weiss and J. Crayton, *Endocrinology*, 87 (1970) 527.
77 L. M. Klainer, Y. M. Chi, S. L. Freidberg, R. W. Rall and E. W. Sutherland, *J. Biol. Chem.*, 247 (1962) 1239.
78 F. Murad, Y. M. Chi, T. W. Rall and E. W. Sutherland, *J. Biol. Chem.*, 237 (1962) 1233.
79 M. H. Makman and E. W. Sutherland, *Endocrinology*, 75 (1964) 127.
80 M. W. Bitensky, J. W. Clancy and E. Gamache, *J. Clin. Invest.*, 46 (1967) 1037.
81 M. W. Bitensky, V. Russell and W. Robertson, *Biochem. Biophys. Res. Comm.*, 31 (1968) 706.
82 M. W. Bitensky, V. Russell and M. Blanco, *Endocrinology*, 86 (1970) 154.
83 S. L. Pohl, L. Birnbaumer and M. Rodbell, *Science*, 164 (1969) 566.
84 G. V. Marinetti, T. K. Ray and V. Tomasi, *Biochem. Biophys. Res. Comm.*, 36 (1969), 185.
85 H. P. Bär and P. Hahn, *Can. J. Biochem.*, 49 (1971) 85.
86 G. A. Robison, R. W. Butcher and E. W. Sutherland, *Ann. N. Y. Acad. Sci.*, 139 (1967) 703.
87 G. S. Levey, *J. Biol. Chem.*, 246 (1971) 7905.
88 F. Murad and M. Vaughan, *Biochem. Pharmacol.*, 18 (1969) 1053.
89 G. S. Levey and S. E. Epstein, *J. Clin. Invest.*, 48 (1969) 1663.
90 J. B. Posner, R. Stern and E. G. Krebs, *J. Biol. Chem.*, 240 (1965) 982.
91 S. E. Mayer and J. T. Struel, *Ann. N. Y. Acad. Sci.*, 1851 (1971) 433.
92 L. Lundholm, T. Rall and N. Vamos, *Acta Physiol. Scand.*, 70 (1967) 127.
93 J. W. Craig, T. W. Rall and J. Larner, *Biochim. Biophys. Acta*, 177 (1969) 213.
94 L. Triner, N. I. A. Overweg and G. G. Nahas, *Nature*, 225 (1970) 282.
95 L. Triner, G. G. Nahas, Y. Vulliemoz, N. I. A. Overweg, M. Verosky, D. V. Habie and S. H. Ngai, *Ann. N. Y. Acad. Sci.*, 185 (1971) 458.
96 S. Kakiuchi and T. W. Rall, *Mol. Pharmacol.*, 4 (1968) 367.
97 G. A. Robison, M. J. Schmidt and E. W. Sutherland, *Advan. Biochem. Psychopharm.*, 3 (1970) 11.
98 H. Shimizu, C. R. Creveling and J. W. Daly, *Advan. Biochem. Psychopharm.*, 3 (1970) 135.
99 T. W. Rall and A. Sattin, *Advan. Biochem. Psychopharm.*, 3 (1970) 113.
100 J. W. Daly, M. Huang and H. Shimizu, in: P. Greengard, R. Paoletti and G. A. Robison (Eds.), *Advan. Cyclic Nucleotide Res.*, Vol. 1, Raven, New York, 1972, p. 121.
101 A. L. Steiner, G. T. Peake, R. D. Utiger, I. E. Karl and D. M. Kipnis, *Endocrinology*, 86 (1970) 354.
102 V. Zor, T. Kaneko, H. P. G. Schneider, S. M. McCann and J. B. Field, *J. Biol. Chem.*,

245 (1970) 2883.
103 J. R. Turtle and D. M. Kipnis, *Biochem. Biophys. Res. Comm.*, 28 (1967) 797.
104 I. D. Goldfine, J. Roth and L. Birnbaumer, *J. Biol. Chem.*, 247 (1972) 1211.
105 H. P. Bär and O. Hechter, *Proc. Natl. Acad. Sci. (U.S.)*, 63 (1969) 350.
106 R. W. Butcher, C. E. Baird and E. W. Sutherland, *J. Biol. Chem.*, 243 (1968) 1705.
107 L. Birnbaumer and M. Rodbell, *J. Biol. Chem.*, 244 (1969) 3477.
108 L. Birnbaumer, S. Pohl and M. Rodbell, *Advan. Biochem. Psychopharm.*, 3 (1970) 185.
109 T. Braun and O. Hechter, *Proc. Natl. Acad. Sci. (U.S.)*, 66 (1970) 995.
110 T. Braun and O. Hechter, in: B. Jeanrenaud and D. Hepp (Eds.), *Adipose Tissue Regulation and Metabolic Functions*, Vol. 1, Thieme, Stuttgart, 1970, p. 11.
111 T. Braun and O. Hechter, in preparation.
112 T. W. Burns and P. E. Langley, *J. Lab. Clin. Med.*, 75 (1970) 983.
113 T. W. Burns, P. E. Langley and G. L. Robison, *Ann. N. Y. Acad. Sci.*, 185 (1971) 115.
114 L. A. Carlson, R. W. Micheli, *Acta Med. Scand.*, 187 (1970) 525.
115 R. W. Butcher and C. E. Baird, in: R. Paoletti (Ed.), *Drugs Affecting Lipid Metabolism*, Plenum, New York, 1969, p. 5.
116 J. Skala, P. Hahn and T. Braun, *Life Sci.*, 9 (1970) 1201.
117 L. R. Chase and G. D. Aurbach, *Science*, 159 (1968) 545.
118 N. Nagata and H. Rasmussen, *Proc. Natl. Acad. Sci. (U.S.)*, 65 (1970) 368.
119 G. L. Melson, L. R. Chase and G. D. Aurbach, *Endocrinology*, 86 (1970) 511.
120 T. Dousa, O. Hechter, I. L. Schwartz and R. Walter, *Proc. Natl. Acad. Sci. (U.S.)*, 68 (1971) 1693.
121 L. R. Chase, S. A. Fedak and G. D. Aurbach, *Endocrinology*, 84 (1969) 761.
122 L. R. Chase and G. D. Aurbach, *J. Biol. Chem.*, 245 (1970) 1520.
123 F. Murad, H. B. Brewer and M. Vaughan, *Proc. Natl. Acad. Sci. (U.S.)*, 65 (1970) 446.
124 J. H. Dorrington and B. Baggett, *Endocrinology*, 84 (1969) 989.
125 J. M. Marsh, *J. Biol. Chem.*, 245 (1970) 1596.
126 I. Murad, B. S. Strauch and M. Vaughan, *Biochim. Biophys. Acta*, 177 (1969) 591.
127 F. A. Kuehl Jr., D. J. Patanelli, J. Tarnoff and J. C. Humes, *Biol. Reprod.*, 2 (1970) 154.
128 M. A. Hollinger, *Life Sci.*, 9 (1970) 533.
129 T. Braun and O. Hechter, *53rd Meeting, Endocrinology Soc.*, 1 (1971) A-97.
130 J. H. Dorrington and I. B. Fritz, *Endocrinology*, 94 (1974) 395.
131 K. J. Catt, K. Watanabe and M. L. Dufau, *Nature*, 239 (1972) 780.
132 M. L. Dufau, K. Watanabe and K. J. Catt, *Endocrinology*, 92 (1973) 6.
133 T. Braun and S. Sepsenwol, *Endocrinology*, 94 (1974) 1028.
134 R. C. Haynes Jr., *J. Biol. Chem.*, 233 (1958) 1220.
135 I. Pastan and R. Katzen, *Biochem. Biophys. Res. Comm.*, 29 (1967) 1792.
136 D. G. Grahame-Smith, R. W. Butcher, R. L. Ney and E. W. Sutherland, *J. Biol. Chem.*, 242 (1967) 5535.
137 A. G. Gilman and T. W. Rall, *J. Biol. Chem.*, 243 (1968) 5867.
138 T. Kaneko, V. Zor and J. B. Field, *Science*, 163 (1969) 1062.
139 O. D. Tauton, J. Roth and I. Pastan, *J. Biol. Chem.*, 244 (1969) 247.
140 R. J. Lefkowitz, J. Roth, W. Pricer and I. Pastan, *Proc. Natl. Acad. Sci. (U.S.)*, 65 (1970) 745.
141 K. Yamashita and J. B. Field, *Biochem. Biophys. Res. Comm.*, 49 (1970) 171.
142 J. Wolff and A. B. Jones, *J. Biol. Chem.*, 246 (1971) 3939.
143 R. P. Ahlquist, *Am. J. Physiol.*, 153 (1948) 586.
144 G. A. Robison and E. W. Sutherland, *Circulation Res.*, 26–27, Suppl. I (1970) I-147.
145 L. Triner, Y. Vulliemoz and G. G. Nahas, *Life Sci.*, 9 (1970) 707.
146 L. S. Jefferson, J. H. Exton, R. W. Butcher, E. W. Sutherland and C. R. Park, *J. Biol. Chem.*, 243 (1968) 1031.

147   S. B. Lewis, J. H. Exton, R. J. Ho and C. R. Park, *Fed. Proc.*, 29 (1970) 379.
148   R. W. Butcher, J. G. T. Sneyd, C. R. Park and E. W. Sutherland, *J. Biol. Chem.*, 241 (1966) 1652.
149   M. Rodbell, A. B. Jones, G. E. Chiappe de Cingolani and L. Birnbaumer, *Recent Progr. Hormone Res.*, 24 (1968) 215.
150   T. K. Ray, V. Tomasi and G. V. Marinetti, *Biochim. Biophys. Acta*, 211 (1970) 20.
151   G. Illiano and P. Cuatracasas, *Science*, 175 (1972) 906.
152   K. D. Hepp, *FEBS Lett.*, 12 (1971) 263.
153   K. D. Hepp, *Europ. J. Biochem.*, 31 (1972) 266.
154   L. Jimenez de Asua, E. Surian, M. M. Flawia and H. N. Torres, *Proc. Natl. Acad. Sci. (U.S.)*, 70 (1973) 1388–1392.
155   T. Braun, H. P. Bär, D. Soifer and O. M. Hechter, *Proc. Meeting of Endocrine Society*, (1969) A-138.
156   P. E. Cryer, L. Jarett and D. M. Kipnis, *Biochim. Biophys. Acta*, 177 (1969) 586.
157   D. O. Allen and R. R. Beck, *Endocrinology*, 91 (1972) 504.
158   B. M. Breckenridge, *Proc. Natl. Acad. Sci. (U.S.)*, 52 (1965) 1580.
159   J. N. Beavo, N. L. Rogers, O. B. Crofford, C. E. Baird, J. G. Hardman, E. W. Sutherland and E. V. Neuman, *Ann. N. Y. Acad. Sci.*, 185 (1971) 129.
160   H. Sheppard, G. Wiggan and W. H. Tsien, in: P. Greengard, P. Paoletti and G. A. Robison (Eds.), *Advances in Cyclic Nucleotide Res.*, Vol. 1, Raven Press, New York, 1972, p. 103.
161   J. W. Kebabian, J. F. Kuo and P. Greengard, in: P. Greengard and G. A. Robison (Eds.), *Advances in Cyclic Nucleotide Res.*, Vol. 2, Raven Press, New York, 1972, p. 131.
162   J. W. Daly, in: M. Chapin (Ed.), *Methods in Cyclic Nucleotide Res.*, Dekker, New York, 1972, p. 255.
163   M. Lin, Y. Solomon and M. Rodbell, *J. Biol. Chem.*, (in press).
164   G. Krishna, B. Weiss and B. B. Brodie, *J. Pharmacol. Exptl. Therap.*, 163 (1968) 379.
165   H. P. Bär and O. Hechter, *Anal. Biochem.*, 29 (1969) 476.
166   L. L. Miller, *Rec. Progr. Hormone Res.*, 17 (1961) 539.
167   J. H. Exton and C. R. Park, *Advan. Enz. Regulation*, 6 (1968) 391.
168   J. H. Exton and C. R. Park, *J. Biol. Chem.*, 243 (1968) 4189.
169   F. A. Robinson, J. H. Exton, C. R. Park and E. W. Sutherland, *Fed. Proc.*, 26 (1967) 257.
170   J. H. Exton, G. A. Robinson, E. W. Sutherland and C. R. Park, *J. Biol. Chem.*, 246 (1971) 6167.
171   T. F. Williams, J. H. Exton, N. Friedmann and C. R. Park, *Am. J. Physiol.*, 221 (1971) 1645.
172   V. Mutt and J. E. Jorpes, *Rec. Progr. Hormone Res.*, 23 (1967) 483.
173   P. W. Felts, M. E. C. Ferguson, K. A. Hagey, E. S. Stitt and W. M. Mitchell, *Diabetologia*, 6 (1970) 44.
174   F. Sundby, in: R. R. Rodriguez, F. J. G. Ebling, I. Henderson and A. Assen (Eds.), *Excerpta Medica ICS*, Vol. 209, Amsterdam, 1970, p. 80.
175   S. L. Pohl, L. Birnbaumer and M. Rodbell, *J. Biol. Chem.*, 246 (1971) 1849.
176   M. Rodbell, L. Birnbaumer, S. L. Pohl and F. Sundby, *Proc. Natl. Acad. Sci. (U.S.)*, 68 (1971) 909.
177   M. Rodbell, L. Birnbaumer, S. L. Pohl and H. M. J. Krans, *J. Biol. Chem.*, 246 (1971) 1877.
178   G. Krishna, J. Harwood, A. J. Barber and G. A. Jamieson, *J. Biol. Chem.*, 247 (1972) 2253.

179 F. Leray, A. M. Chambaut and J. Janoune, *Biochem. Biophys. Res. Comm.*, 48 (1972) 1385.
180 J. Wolf and G. H. Cook, *J. Biol. Chem.*, 248 (1973) 350.
181 M. Rodbell, H. M. J. Krans, S. L. Pohl and L. Birnbaumer, *J. Biol. Chem.*, 247 (1971) 1861.
182 P. Cuatrecasas, *Proc. Natl. Acad. Sci. (U.S.)*, 68 (1971) 1264.
183 B. J. Campbell, G. Woodward and B. Borberg, *J. Biol. Chem.*, 247 (1972) 6167.
184 R. J. Lefkowitz, G. W. G. Sharp and E. Haber, *J. Biol. Chem.*, 248 (1973) 342.
185 C. Y. Lee and R. J. Ryan, *Biochemistry*, 12 (1973) 4609.
186 J. Bockaert, Ch. Roy, R. Rayerison and S. Jard, *J. Biol. Chem.*, 248 (1973) 5923.
187 L. Birnbaumer, S. L. Pohl, M. Rodbell and F. Sundby, *J. Biol. Chem.*, 247 (1972) 2038.
188 L. Birnbaumer, in: M. Margoulies and F. C. Greenwood (Eds.), *Structure–Activity Relationships of Protein and Polypeptide Hormones*, Excerpta Medica ICS, Vol. 241, Amsterdam, 1972, p. 471.
189 S. L. Pohl, H. M. J. Krans, L. Birnbaumer and M. Rodbell, *J. Biol. Chem.*, 247 (1972) 2295.
190 M. W. Bitensky, R. E. Gorman, A. H. Neufeld and R. King, *Endocrinology*, 89 (1971) 1242.
191 M. Rodbell, H. M. J. Krans, S. L. Pohl and L. Birnbaumer, *J. Biol. Chem.*, 246 (1971) 1872.
192 S. L. Pohl, H. M. J. Krans, V. Kozyreff, L. Birnbaumer and M. Rodbell, *J. Biol. Chem.*, 246 (1971) 4447.
193 L. Birnbaumer and S. L. Pohl, *J. Biol. Chem.*, 248 (1973) 2056.
194 G. S. Levey, *Biochem. Biophys. Res. Comm.*, 42 (1971) 103.
195 R. G. Yount, D. Babcock, W. Ballantyne and D. Ojala, *Biochemistry*, 10 (1971) 2484.
196 A. Sattin and R. W. Rall, *Mol. Pharmacol.*, 6 (1970) 13.
197 J. N. Fain, R. H. Pointer and W. F. Ward, *J. Biol. Chem.*, 247 (1972) 6866.
198 J. P. Perkins and M. M. Moore, *J. Biol. Chem.*, 246 (1971) 62.
199 J. P. Perkins, in: P. Greengard and A. G. Robison (Eds.), *Advances in Cyclic Nucleotide Research*, Vol. 3, Raven Press, New York, 1973.
200 L. Birnbaumer, *Biochim. Biophys. Acta*, 300 (1973) 129.

Subject Index